GUIDELINES FOR
Vapor Release Mitigation

GUIDELINES FOR

Vapor Release Mitigation

Prepared by
Richard W. Prugh
Center for Chemical Process Safety
and
Robert W. Johnson
Battelle Columbus Division

for

CENTER FOR CHEMICAL PROCESS SAFETY
of the
American Institute of Chemical Engineers
345 East 47th Street, New York, NY 10017

Library of Congress Cataloging-in-Publication Data

Prugh, Richard W.
    Guidelines for vapor release mitigation.

    Bibliography: p.
    Includes index.
1. Chemical plants — Environmental aspects.　2. Petro-
leum chemicals industry — Environmental aspects.
3. Vapors — Environmental aspects.　4. Hazardous
substances — Environmental aspects.　　I. Johnson, Robert W.
(Robert William), 1955–　　II. American Institute of
Chemical Engineers. Center for Chemical Process Safety.
III. Title.
TD888.C5P78　　1987　　　660.2'804　　　87-26987
ISBN 0–8169–0401–4

**This book is available at a special discount when ordered
in bulk quantities. For information, contact the Center
for Chemical Process Safety at the address shown above.**

It is sincerely hoped that the information presented in this document will
lead to an even more impressive safety record for the entire industry;
however, neither the American Institute of Chemical Engineers, nor the Battelle
Memorial Institute can accept any legal liability or responsibility whatsoever
for the consequences of its use or misuse by anyone.

CONTENTS

Preface
Acknowledgments
Summary
Glossary

1. **Introduction**
    1.1 Objective
    1.2 Hazard of Accidental Vapor Cloud Releases
    1.3 Types of Vapor Clouds
        1.3.1 Flammable Vapor Clouds
        1.3.2 Toxic Vapor Clouds
        1.3.3 Flammable-Toxic Vapor Clouds
        1.3.4 Other Types of Vapor Clouds
    1.4 Forms of Vapor Release
    1.5 Release Causes
    1.6 Possible Consequences of Vapor Cloud Releases
        1.6.1 Toxic Effects
        1.6.2 Fires
        1.6.3 Explosions
    1.7 Analysis of the Need for Mitigation
    1.8 Vapor Release Mitigation Approaches

2. **Mitigation through Inherently Safer Plants**
    2.1 Inventory Reduction
    2.2 Chemical Substitution
    2.3 Process Modification
        2.3.1 Refrigerated Storage

2.3.2 Dilution 19
2.4 Siting Considerations 20

3. Engineering Design Approaches to Mitigation 23
3.1 Plant Physical Integrity 24
   3.1.1 Design Practices 24
   3.1.2 Materials of Construction 25
3.2 Process Integrity 27
   3.2.1 Identification of Reactants and Solvents 27
   3.2.2 Limits on Operating Conditions 28
   3.2.3 Process Control Systems 29
   3.2.4 Pressure Relief Systems 30
      3.2.4.1 Relief Devices 30
      3.2.4.2 Relief Headers 31
3.3 Process Design Features for Emergency Control 33
   3.3.1 Emergency Relief Treatment Systems 33
      3.3.1.1 Active Scrubbers. 34
      3.3.1.2 Passive Scrubbers 36
      3.3.1.3 Stacks 37
      3.3.1.4 Flares 40
      3.3.1.5 Catchtanks for Vapor-Liquid Separation 43
      3.3.1.6 Incinerators 45
      3.3.1.7 Absorbers 46
      3.3.1.8 Adsorbers 46
      3.3.1.9 Condensers 47
   3.3.2 Emergency Process Abort Systems 47
   3.3.3 Emergency Isolation of Leak/Break 49
      3.3.3.1 Isolation Devices 50
      3.3.3.2 Remote Isolation. 52
      3.3.3.3 Inspection and Testing of Isolation
             Devices 52
   3.3.4 Emergency Transfer of Materials 53
      3.3.4.1 Transfer of Vapor/Cover Gas to Reduce
             Driving Pressure 54
      3.3.4.2 Transfer of Liquids to Reduce Inventory
             Available for Release 56
3.4 Spill Containment 57
   3.4.1 Double Containment 57
   3.4.2 Enclosures and Walls 57
   3.4.3 Dikes, Curbs, Trenches, and Impoundments 58

4. Process Safety Management Approaches to Mitigation 63
4.1 Operating Policies and Procedures 63
4.2 Training for Vapor Release Prevention and Control 66

4.3 Audits and Inspections   67
4.4 Equipment Testing   68
4.5 Maintenance Programs   70
4.6 Modifications and Changes   71
4.7 Methods for Stopping a Leak   73
     4.7.1 Patching   73
     4.7.2 Freezing   74
4.8 Security   75

5. Mitigation through Early Vapor Detection and Warning   77
5.1 Detectors and Sensors   77
     5.1.1 Types of Sensors   78
            5.1.1.1 Combustion   78
            5.1.1.2 Catalytic   78
            5.1.1.3 Electrical   78
            5.1.1.4 Chemical Reaction   78
            5.1.1.5 Visual   79
            5.1.1.6 Absorption/Scattering   79
     5.1.2 Response Time of Sensors   79
     5.1.3 Positioning of Sensors   81
5.2 Detection by Personnel   81
     5.2.1 Odor Warning Properties   82
     5.2.2 Color or Fog   82
5.3 Alarm Systems   84

6. Mitigation through Countermeasures   87
6.1 Vapor/Liquid Releases   87
6.2 Vapor Release Countermeasures   88
     6.2.1 Water Sprays   88
     6.2.2 Water Curtains   89
     6.2.3 Steam Curtains   90
     6.2.4 Air Curtains   91
     6.2.5 Deliberate Ignition   91
     6.2.6 Ignition Source Control   92
6.3 Liquid Release Countermeasures   93
     6.3.1 Dilution   94
     6.3.2 Neutralization   94
     6.3.3 Covers   94
            6.3.3.1 Liquids   94
            6.3.3.2 Foams   95
            6.3.3.3 Solids   96
            6.3.3.4 Application   96
6.4 Avoidance of Factors that Aggravate Vaporization   96

7. On-Site Emergency Response   **101**
   7.1 On-Site Communications   102
   7.2 Emergency Shutdown Equipment and Procedures   103
   7.3 Site Evacuation   104
   7.4 Havens   104
   7.5 Escape from Vapor Cloud   106
   7.6 Personal Protective Equipment   106
   7.7 Medical Treatment   108
   7.8 On-Site Emergency Plans, Procedures, Training,
        and Drills   108

8. Alerting Local Authorities and the Public   **113**
   8.1 Alerting Systems   114
        8.1.1 Capabilities   114
        8.1.2 Input requirements   115
   8.2 Roles and Lines of Communication   115
   8.3 Information to Be Communicated   116

9. Selection of Mitigation Measures   **117**
   9.1 Risk Analysis   118
   9.2 Methods for Hazard Identification   120
   9.3 Methods for Estimating the Consequences of Accidents   120
   9.4 Methods for Estimating the Probability of Accidents   122

**Appendix A. Loss-of-Containment Causes in the
               Chemical Industry**   **123**

Appendix B. Properties of Some Hazardous Materials.   **129**

Appendix C. Derivation of Fog Correlations   **131**

Appendix D. Catchtank Design   **133**

Appendix E. Capacity of Havens   **137**

Appendix F. Sources to Vapor-Mitigation Equipment Vendors   **141**

Subject Index   **143**

PREFACE

The American Institute of Chemical Engineers (AIChE) has a 30-year history of involvement with process safety and loss control for chemical and petrochemical plants. Through its strong ties with process designers, plant builders and operators, safety professionals, and academia, the AIChE has enhanced communication and fostered improvement in the high safety standards of the industry. Their publications and symposia have become an information resource for the engineering profession on the causes of accidents and means of prevention.

Early in 1985, AIChE established the Center for Chemical Process Safety (CCPS) to serve as a focus for a continuing program for process safety. The first CCPS project was the preparation of *Guidelines for Hazard Evaluation Procedures*. One of the CCPS projects for 1987 was the preparation of this document, *Guidelines for Vapor Release Mitigation*. The goal of this project was to

> publish available information on generic techniques designed to reduce the consequences of unplanned hazardous vapor releases. Sources of information will be major chemical companies as well as recent open literature, governmental agencies, transport systems, and engineering organizations. The CCPS will solicit information from major chemical companies and by so doing provide these companies a mechanism for making any special knowledge available to the engineering community and the public.

Thus, *Guidelines for Vapor Release Mitigation* is a survey of current industrial practice for controlling accidental releases of hazardous vapors and preventing their escape from the source area. To prepare this document, CCPS reviewed the available literature for descriptions of existing and proposed vapor-control equipment and visited industrial sites. CCPS also obtained equipment designs and procedures for dealing with vapor releases from chemical and petrochemical companies.

These guidelines are intended to represent current industrial practice rather than theory. However, some of the suggested practices and equipment have not been fully tested; that is, they may not have been used to mitigate an actual vapor release. These guidelines present methods for attaining improvement and are a starting point for further development; however, they are not proposed as standards to be achieved by the industry, and companies are not expected to employ all of the methods presented.

Further, there are wide variations in toxicity (acute, chronic, and latent), resistance to corrosion and erosion, flammability (explosive limits and ignition energies), and physical properties (vapor pressure and vapor density) among the fluids involved. Therefore, the applicability of the guidelines should be evaluated or tested in terms of particular fluids and proposed construction materials.

Guidelines for Vapor Release Mitigation should be useful to both experienced and inexperienced engineers, but because of the rapid evolution in plant design and operation, it is unlikely that this volume includes all the useful methods for mitigating vapor hazards. Current literature--particularly the journals of AIChE and its British counterpart, the Institution of Chemical Engineers--contains additional guidance on existing and novel methods of vapor control.

Eliminating the cause of releases and reducing their frequency are effective mitigation methods, in the broader sense of the term. Techniques for analyzing processes to identify vapor release sources and evaluate their likelihood are presented in *Guidelines for Hazard Evaluation Procedures* and in a forthcoming CCPS volume, *Guidelines for Chemical Process Quantitative Risk Assessment Procedures*. Methods for preventing vapor releases are also addressed in *Guidelines for Safe Storage and Handling of High Toxic Hazard Material*. Although methods for reducing the likelihood of vapor releases are included in the present volume, these guidelines emphasize methods for reducing the size, duration, and consequences of vapor releases.

ACKNOWLEDGMENTS

The AIChE wishes to thank the members of the Technical Steering Committee of the Center for Chemical Process Safety for their advice and support. Under the auspices of the Technical Steering Committee, the Vapor Mitigation Subcommittee of the CCPS provided guidance in this work. The chairman of the subcommittee was Stanley J. Schechter (Rohm and Haas), with Edwin J. Bassler (Stone & Webster) and G. A. Viera (Union Carbide) also on the subcommittee. Thomas W. Carmody and Russell G. Hill of the Center for Chemical Process Safety were responsible for overall administration and coordination of this project.

The subcommittee acknowledges the assistance of the following people in commenting on various aspects of the drafts: D. E. Wade (Monsanto), R. A. Smith (Dow), R. F. Schwab (Allied-Signal), R. G. Holmes (Westinghouse), R. W. Ormsby (Air Products), P. Rasch (Celanese), R. J. Hawkins (Celanese), S. S. Grossel (Hoffman-LaRoche), and J. Hagopian (Arthur D. Little).

The principal author of *Guidelines for Vapor Release Mitigation* was Richard W. Prugh, a staff member of the Center for Chemical Process Safety, with significant technical and editorial contributions by Robert W. Johnson of Battelle Memorial Institute's Columbus Division. Important editorial contributions from William H. Goldthwaite of Battelle Columbus Division are gratefully acknowledged.

SUMMARY

The purpose of this document is to make generally available the approaches and measures that are currently being used by many companies in the chemical process industry for mitigating the likelihood and consequences of vapor cloud releases.

Many of the major chemical accidents that have occurred recently have involved the release of toxic or flammable vapors in quantities sufficient to have severe health and environmental impacts. The release of dangerous amounts of toxic and flammable vapors can be minimized and the severity of their effects can be reduced by a variety of mitigation measures. The choice of mitigation measure depends upon the particular hazard of concern, the amount of material involved, the siting of the facility, the processes involved, and other characteristics of the facility in question. Nevertheless, there is a hierarchy or preference order to the approaches that should be considered in choosing an approach for any particular mitigation concern.

The mitigation approach which is generally most effective is to make the plant inherently safer. A chemical process facility will be inherently safer if, for example, the inventory of hazardous material can be eliminated by substitution of a nonhazardous material in the process, or reduced to a level where total release would not pose a threat to employees or the public. This is being done in several facilities by manufacturing the hazardous material in situ and limiting the inventory to that which is in the pipes and reaction vessels. Another approach that is sometimes possible with new plants is to choose a site far enough removed from populated or sensitive areas that it would be impossible for a hazardous concentration to develop there. It follows that this buffer zone must somehow be kept inviolate for the life of the plant or process.

There are a variety of engineering approaches to mitigation of hazardous vapor releases that are next in order of preference. The first is to ensure plant integrity so that the probability of a release is minimized. Attention to design and construction codes is an important aspect of this. Ensuring that materials of construction are chosen to maintain plant integrity while containing the process materials under the process conditions and under process upset conditions is essential. Inspection and testing of materials and equipment before start-up and at intervals during operation of the process are also necessary.

"Process integrity" is also high on the preference order of approaches to mitigation. Process integrity involves the chemistry of plant design and operation, and includes ensuring that only the proper reactants and solvents are used and that they are of the required purity for the process and equipment. Knowing and maintaining the conditions of operation within limits that are known to be safe is also essential to process integrity. Avoiding processes that are sensitive to parameter deviations and providing measurement and control of those parameters are helpful in avoiding upsets that lead to loss-of-containment accidents. Pressure relief systems, properly chosen and configured and venting into other containment or disposal systems such as scrubbers, stacks, and flares, are another way of enhancing process integrity.

The approaches to mitigation that have been mentioned above are at the top of the preference order because they can work toward preventing the release of dangerous amounts of hazardous vapors. However, mitigation after loss of containment can also be effective and usually must be provided for in the process design stage. Secondary containment by concentric piping, double-walled vessels, or enclosures may be warranted. In the event of the release of a volatile liquid to the environment by a leak or a rupture of a line or vessel, containing the liquid in a restricted area or minimizing its surface area can reduce the quantity of material released as vapor. This can be accomplished by using dikes, curbs, and trenches leading to strategically located impoundments. An added incentive is to help keep the liquid source of the vapor away from sensitive areas with respect to people, process, and the environment.

There are other effective mitigation measures that are more concerned with the start-up, operation, and maintenance of the facility than with facility and process design. These include procedure development and communication, training, inspections and tests, documentation and the protocol for maintenance and modifications. In each instance, accuracy and clarity of procedures are important to avoid errors which could lead to upsets and vapor release accidents.

Should loss of containment develop, there are methods for temporarily stopping a leak through patching or, with some materials, freezing that may be useful. Mitigation by early detection and warning can be effective in preventing on-site health effects and in limiting the off-site release to nonhazardous levels. There are several detection methods ranging from various types of sensors to detection by personnel by odor or sight. Warning may involve alarms, communication systems, and accident analysis systems.

Lower on the preference order, but still important, are the systems or equipment that can be provided for mitigation by countermeasures. These provide mitigation by controlling, to as great a degree as possible, the dilution and dispersion of the hazardous vapors. Systems such as water curtains, steam curtains and water sprays have been studied and used, to a limited extent, to control the movement and concentration of the vapor. If the spilled liquid has been confined by a dike or impoundment, reducing the vapor generation rate can often be done by covering the liquid source with foams, with compatible liquids, and, in some cases, with granular solids.

On-site and off-site emergency response can provide effective mitigation of health effects in many instances if adequately planned and effectively implemented. This requires both equipment and training--both tailored to the materials and types of accidents that are most likely to be involved.

The preference order of mitigation measures that have been mentioned and are described in this volume is important to selection of the most appropriate measures for a particular facility. Also important is an understanding of the types of accident that can occur and result in a vapor release and the possible consequences of such a release. Methods for identifying and evaluating accident scenarios, both with and without particular mitigation measures, can be valuable in the selection process. Other AIChE-CCPS volumes, referenced in this text, describe these identification and evaluation procedures.

GLOSSARY

Accident, Accident Event Sequence
A specific unplanned event or sequence of events that has a specific undesirable consequence.

Acute Exposure
A short-term or rare exposure to a toxic agent in a single episode which is unlikely to recur.

Acute Hazard
The potential for injury or damage to occur as a result of an instantaneous or short duration exposure to the effects of an undesirable event (e.g., an explosion with the potential for causing damage and injury).

Audit (Process Safety Audit)
An inspection of a plant/process unit, drawings, procedures, emergency plan and/or management systems, etc., usually by an off-site team. (See "Review" for contrast.)

Boiling-Liquid-Expanding-Vapor Explosion (BLEVE)
A type of rapid phase transition in which a liquid contained above its atmospheric boiling point is rapidly depressurized, causing a nearly instantaneous transition from liquid to vapor with a corresponding energy release. A BLEVE is often accompanied by a large fireball if a flammable liquid is involved, since an external fire impinging on the vapor space of a pressure vessel is a common BLEVE scenario. However, it is

not necessary for the liquid to be flammable to have a BLEVE occur.

Chronic Exposure   A frequent or continuous exposure to a toxic agent over an unspecified but generally lengthy period of time.

Chronic Hazard   The potential for injury or damage to occur as a result of prolonged exposure to an undesirable condition (e.g., smoking with the potential for causing lung cancer).

Consequence   The result of an accident event sequence. In this document it is the fire, explosion, release of toxic material, etc., that results from the accident but not the health effects, economic loss, etc., that is the ultimate result.

Emergency   One-hour concentrations being developed
Response   for a limited number of industrial chemicals
Planning   having high vapor toxicity properties, for use by
Guidelines   industry as guidelines in emergency planning
(ERPG)   and response activities.

Event   An occurrence involving equipment performance or human action, or an occurrence external to the system that causes system upset. In this document an event is associated with an accident either as the cause or a contributing cause of the accident or as a response to the accident-initiating event.

External Event   An occurrence external to the system/plant such as an earthquake, a flood, or an interruption of facilities such as electric power or process air.

Hazard   A characteristic of the system/plant/process that represents a potential for an accident. In this document, it is the combination of a (hazardous) material and an operating environment such that certain unplanned events could result in an accident.

Immediately Dangerous to Life and Health (IDLH)
: Maximum airborne contaminant concentrations from which one could escape within 30 minutes without any escape-impairing symptoms or any irreversible health effects. Developed by the National Institute for Occupational Safety and Health (NIOSH).

Initiating Event
: An event that will result in an accident unless systems or operations intervene to prevent or mitigate the accident.

Intermediate Event
: An event in an accident event sequence that helps to propagate the accident or helps to prevent the accident or mitigate the consequences.

Mitigation
: As used in this document, lessening the risk of an accident event sequence by acting on the source in a preventive way by reducing the likelihood of occurrence of the event, or in a protective way by reducing the magnitude of the event and/or the exposure of local persons or property. The emphasis in this document is "mitigating" or reducing the size, duration, and consequences of vapor cloud releases.

Mitigation System
: Equipment and/or procedures designed to respond to an accident event sequence by hindering accident propagation and/or reducing the accident consequences.

Primary Event
: A basic independent event for which frequency can be obtained from experience or test.

Probability
: An expression for the likelihood of occurrence of an event or an event sequence during an interval of time, or the likelihood of failure of a component in response to a test or demand.

Review (Process Safety Review)
: An inspection of a plant/process unit, drawings, procedures, emergency plans and/or management systems, etc., usually by an on-site team and usually problem-solving in nature. (See "Audit" for contrast).

Risk

A measure of potential economic loss or human injury in terms of the probability of the loss or injury occurring and the magnitude of the loss or injury if it occurs.

Safety System

Equipment and/or procedures designed to respond to an accident event sequence by preventing accident propagation, thereby preventing the accident and its consequences.

Toxic Hazard

In the context of these guidelines, a measure of the danger posed to living organisms by a toxic agent, determined not only by the toxicity of the agent itself but also by the means by which it may be introduced into the subject organisms under prevailing conditions.

Toxicity

The quality, state, or degree to which a substance is poisonous and/or may chemically produce an injurious or deadly effect upon introduction into a living organism.

"Worst Case" Consequence

A conservative (high) estimate of the consequences of the most severe accident identified. For example, the assumption that the entire contents of a contained volume of toxic material is released to the most vulnerable area in such a way (all at once or continuous) as to have the maximum effect on the public or employees in that area. The contained volume could be chosen as the containers and pipes between shutoff valves or the entire process unit but probably not the entire plant.

GUIDELINES FOR
Vapor Release Mitigation

1

INTRODUCTION

The preparation of this document was sponsored by the Center for Chemical Process Safety of the American Institute of Chemical Engineers (AIChE). The Center for Chemical Process Safety (CCPS) was established by the AIChE early in 1985 as a focal point to further the development and communication of improved safety methods, practices, designs, and procedures in the chemical process industry. One of the goals of the CCPS has been to prepare a series of books which cover different aspects of process safety in order to provide a means of sharing approaches that have proved to be successful with the rest of the industry. The subject matter of these volumes ranges from procedures for identifying and evaluating hazards to safe ways of storing and handling highly toxic materials.

It is significant that many of the major chemical accidents that have occurred in the past decade or so have involved the release and transport of toxic or flammable vapors to populated or environmentally sensitive areas. This volume addresses the vapor release concern by describing a variety of methods and measures that can be used to control the risk of hazardous vapor releases.

Line drawings have been placed throughout the text to help the reader visualize the concepts involved in the various mitigation approaches. These are not intended to be accurate engineering drawings, but rather more like caricatures of selected methods. A "base case" drawing of the process which is used to convey these concepts is presented as follows:

Base Case

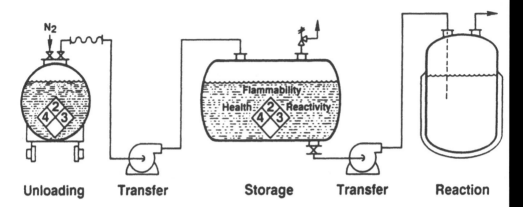

| Unloading | Transfer | Storage | Transfer | Reaction |

This drawing implies the unloading of a toxic and flammable material from a rail car to a large on-site storage tank via an unloading line having a flexible connection, and the subsequent pumping of the material into a liquid-phase reactor vessel. In this example it is assumed that the material is converted in the reactor into a less hazardous product, which is purified in downstream equipment.

1.1 OBJECTIVE

The objective of these engineering guidelines is to provide information to designers and operators of process plants on a variety of measures which can be used for mitigating the effects of unplanned vapor release incidents. In these guidelines, "mitigation" is defined as lessening the risk of a release incident by acting on the source (at the point of release) either (1) in a preventive way by reducing the likelihood of an event which could generate a hazardous vapor cloud or (2) in a protective way by reducing the magnitude of the release and/or the exposure of local persons or property. Thus, the term "mitigation" is used quite broadly in these guidelines. Since another AIChE-CCPS document entitled *Guidelines for Safe Storage and Handling of High Toxic Hazard Materials* addresses various plant design and construction considerations, the emphasis in this document will be "mitigation" in the more specific sense of reducing the size, duration, and/or consequences of vapor cloud releases.

1.2 HAZARD OF ACCIDENTAL VAPOR CLOUD RELEASES

Normally, the integrity of process equipment and containment of process materials are not problems in chemical plants. Occasionally, however, process fluids leak from equipment, usually as a result of wear and tear, and site personnel respond by stopping minor releases via adjustment, maintenance, or repair. More serious incidents occur infrequently and unexpectedly. In these cases, the response is to take prompt action to prevent exposure of personnel and the public. The incident is then investigated, and equipment or procedures are modified as required to prevent recurrence.

Recent incidents have demonstrated that major releases of toxic or flammable fluids can sometimes cause injury or property losses that are orders of magnitude greater than the costs of improved design, operation, and equipment testing. The chemical and petrochemical industries are taking action to prevent such incidents by identifying the hazards inherent in handling the material, by safeguarding against these hazards in design and construction of new plants, and by periodic review of existing processes. Hazards identification and review methods are presented in *Guidelines for Hazard Evaluation Procedures* (AIChE-CCPS, 1985). In addition, local and federal agencies are developing legislation directed toward minimizing such incidents, particularly those with potential for causing injury to the public (Baer and Kean, 1986; Waxman et al., 1985; United Kingdom, 1975; Council of European Economic Communities, 1982; CERCLA, 1986).

1.3 TYPES OF VAPOR CLOUDS

Gases and vapors emitted from equipment as well as vapors resulting from evaporation of mists or spilled liquids can form vapor clouds having various properties such as flammability, toxicity, visibility, and odor. The term "vapor" as used in these guidelines is used in the general sense of any substance in the gaseous state. The term "cloud" indicates that the vapors are essentially unconfined in the atmosphere; i.e., they are not inside a building or enclosure.

1.3.1 Flammable Vapor Clouds

Flammable vapor clouds are those which involve vapors of a flammable or combustible material. To pose a flammability hazard, the vapor concentration, at least in a localized pocket, must be above the vapor's lower flammable limit. (Although vapors in concentrations above the vapor's upper flammable limit are not technically flammable, in practice any vapors in a vapor cloud above the lower or upper flammable limit are considered hazardous since turbulence and further dilution by dispersion can entrain enough air to allow combustion to occur.)

1.3.2 Toxic Vapor Clouds

Toxic vapor clouds are those which involve vapors of a material with acute toxicity properties, generally by inhalation. Long-term, continuous, low-level releases (such as fugitive emissions) and chronic health hazards are not addressed in these guidelines.

1.3.3 Flammable-Toxic Vapor Clouds

Several materials, such as hydrogen sulfide, hydrogen cyanide, methyl isocyanate, and acrolein, are both flammable and toxic. This combination compounds the hazard but may provide an opportunity for using unique methods of mitigation, such as deliberate ignition of the released material.

1.3.4 Other Types of Vapor Clouds

Besides flammability and inhalation toxicity, vapor clouds can have other undesirable characteristics such as odor and visibility. Although not all are hazards per se, these characteristics may generate complaints from surrounding areas and result in a less-than-desirable workplace. Drifting of dense vapor clouds across highways has resulted in traffic accidents with associated injuries. Vapor clouds of substances such as hydrogen fluoride can cause chemical burns to the skin, and absorbtion of some toxic substances through the skin can cause systemic injuries. Some vapors may cause property losses, for example by damaging painted surfaces. Although these guidelines are directed toward flammable and toxic vapor clouds, many of the mitigation techniques apply to these other types of vapor clouds as well.

1.4 FORMS OF VAPOR RELEASE

Release of a hazardous material can create a cloud near the source of release that may extend downwind (or downhill, for dense vapors and low wind speeds). The fraction of the released material that forms a vapor cloud varies with the properties of the material:

- *Noncondensing vapor.* If a substance that has a boiling point well below the ambient temperature is released in vapor form, the cloud will contain all of the material, diluted with air.

- *Condensing vapor.* If a substance having a boiling point above ambient temperature is released as a vapor, some of the vapor may condense. Some of the condensate may drop out of the cloud (and vaporize later), and the remainder may stay in the cloud as aerosol or mist.
- *Two-phase flow.* If a flashing liquid (i.e., a liquid stored or processed above its boiling point) or a two-phase, vapor-liquid mixture is released, the cloud is likely to contain liquid droplets, aerosol, fog, or mist in addition to the vapor.
- *Nonflashing liquid.* If liquid is released at a temperature below its boiling point, forming a pool on the ground, the cloud will contain the vapors emitted from the pool, which may be only a small fraction of the spill in quantity. Some aerosol formation may occur if it is a pressurized release.

Special Considerations for Flashing Liquid Releases. Many of the most severe incidents following release of hazardous vapors have involved flashing liquid releases. Very large vapor clouds can be formed when loss of containment occurs in processes handling liquids stored or processed above their normal boiling points. For this reason, special attention has been focused upon analysis of such situations. In particular, the following should be noted:

- *Adiabatic flashing.* A fraction of the liquid immediately flashes to vapor, with the flashing fraction generally calculated assuming adiabatic conditions. The higher the temperature is above the normal boiling point of the liquid, the greater the flashing fraction is, with the fraction rapidly nearing unity as the critical temperature of the liquid is approached.
- *Mist formation.* As mentioned above, the cloud is likely to contain liquid droplets, aerosol, fog, or mist in addition to the material which has flashed to vapor. Since the flashing liquid release is usually a pressurized release, the fraction so released can be substantial. Recent studies have attempted to quantify the mist fraction (IChemE, 1986). This aerosol/mist can evaporate from heat input from the surrounding air and solar radiation as the cloud drifts downwind, and thus contribute to the mass of material in the vapor phase.
- *Pool boiling.* Since flashing liquid releases often involve materials with boiling points below ambient temperature (such as ammonia and hydrogen chloride), any released material contacting the ground as a liquid is subject to pool boiling, because the ground surface provides heat for vaporization.

- *Mass flow rate.* Release of flashing liquids may lead to larger clouds than could result from an all-vapor release through the same size breach of containment, because the increased mass flow may more than compensate for the fraction of liquid which does not flash, become entrained in the cloud, or subsequently boil off or vaporize.
- *Flow restriction.* Release of a flashing liquid through an orifice under its own vapor pressure can be significantly restricted by the flashing. A nonequilibrium flow model must be used to calculate the leak rate if the nozzle length is less than about 4 inches (Fauske, 1985). The calculations can be further complicated by the presence of a pad pressure in the vessel; a fluid flow expert should generally be consulted for such calculations.
- *Jet entrainment.* Pressurized release of a flashing liquid will draw air into the developing plume due to the momentum of the material being released as a jet. This additional air entering into the cloud provides an additional heat source for vaporization of entrained mist, and needs to be considered for an accurate assessment of the source term when performing dispersion calculations.
- *Dense vapors.* The adiabatic flashing of a liquid decreases the temperature of the material to its boiling point. Further vaporization of the liquid and/or mist also results in cooling of the material. When the material's boiling point is below ambient temperature, the vapors in the cloud will be cold and thus are more likely to be denser than the surrounding air. This also needs to be considered in dispersion modeling.
- *Low-temperature embrittlement.* The cooling effect of the adiabatic flashing will cause the process equipment to become cold. This may need to be considered when selecting materials of construction. Low-temperature embrittlement could cause further equipment damage and worsening of the leak.
- *Freezing of material.* The cooling of the process material has also been known to reduce the temperature to its freezing point, particularly for materials where the boiling point and freezing point are relatively close. In some cases, this has resulted in substantial reduction or even stoppage of a small leak; however, this phenomenon should not be relied upon for leak protection.

1.5 RELEASE CAUSES

Nearly all vapor release events are caused by loss of containment of a volatile material. However, toxic clouds can result from releases of less toxic materials (1) by the burning of some materials or mixtures which results in the generation of toxic combustion products, such as the burning of chlorinated organics producing hydrogen chloride; and (2) by the chemical reaction of a spilled material with water or some other commonly present material, such as the reaction of chlorosulfonic acid with humidity in the air to produce sulfuric acid/ hydrochloric acid mist. Toxic clouds can also be formed by the vaporization of a low-volatility liquid from a fire involving a mixture of the toxic material with a combustible substance, such as the burning of a PCB/oil mixture (with the possible formation of dioxins in this case).

Various system failure events can result in *loss of containment* of process material. Examples of such events are

- overpressuring a process or storage vessel due to loss of control of reactive materials or external heat input;
- overfilling of a vessel or knock-out drum;
- opening of a maintenance connection during operation;
- major leak at pump seals, valve stem packings, flange gaskets, etc.;
- excess vapor flow into a vent or vapor disposal system;
- tube rupture in a heat exchanger;
- fracture of a process vessel causing sudden release of the vessel contents;
- line rupture in a process piping system;
- failure of a vessel nozzle;
- breaking off of a small-bore pipe such as an instrument connection or branch line; and
- inadvertently leaving a drain or vent valve open.

The causes of these types of loss-of-containment events can generally be divided into four categories: (1) "open-end" routes to the atmosphere, (2) imperfections in, or deterioration of, equipment integrity, (3) external impact, and (4) deviations from design conditions. A more detailed breakdown of loss-of-containment causes in each category is given in Appendix A in outline form. Although this "Loss-of-Containment Checklist" cannot be taken as exhaustive, it may be helpful to plant engineers in identifying weaknesses or areas needing more careful analysis, including use as part of the hazards analysis methods listed in chapter 9.

1.6 POSSIBLE CONSEQUENCES OF VAPOR CLOUD RELEASES

The consequences of accidental vapor cloud releases having the greatest severity in terms of acute health hazards are usually toxic releases, fires, and explosions. The actual consequences in any particular situation will depend on a number of factors, including the nature and extent of the cloud, the physical and environmental situation at the time of the release, and the location of people with respect to release and ignition locations.

1.6.1 Toxic Effects

The consequences of a toxic vapor release result from the dispersed concentration of vapors, the toxicity of the material, the duration of exposure, the vulnerability of potentially exposed populations (e.g., persons who are very young, very old, immobile, or have a preexisting health condition), and any self-protection measures which may be effective. The severity of consequences of a toxic vapor release is highly dependent on the dispersion of the vapors before reaching persons downwind of the release. The dispersion is a function of many variables: distance, wind speed, release height, release direction and velocity, atmospheric turbulence, solar radiation, roughness and other features of the terrain, vapor density, and interaction with, for instance, moisture in the air.

Examples of methods for mitigating the effects of toxic vapor clouds include limiting the amount released and/or the rate of release, reducing the vaporization of liquid pools by covering, enhancing the dispersion of the released vapor, providing effective personal protective equipment to on-site personnel, evacuating or providing havens for potentially exposed persons, and providing medical treatment if necessary. The most effective measure to consider is substitution of the process material of concern with a less toxic material, if feasible.

1.6.2 Fires

Flammable vapor clouds can result in fires, fireballs, flash fires, or explosions. (These combustion terms are not precise in meaning and are sometimes used interchangeably.) Immediate ignition of a continuous vapor release generally leads to a *fire*, which may be highly directional in a torchlike manner. In this case, the fire intensity depends on the rate of release, the heat of combustion, and the luminosity of the flame. The immediate ignition of an instantaneous ("puff") release of volatile flammable material, as from a bursting container or boiling-liquid-expanding-vapor explosion (BLEVE), will result in a *fireball*. The size of the fireball depends on the total quantity of flammable vapor released (plus flammable/combustible mist, if any) and the heat of combustion. Delayed ignition of a flammable vapor cloud where no vapor cloud explosion results (as described in the next section) will cause a *flash fire*, with similar considerations as a fireball.

Fire hazards posed by the generation of flammable vapor clouds can be mitigated by limiting the amount of material released (quantity, duration), limiting the rate of release, dispersing the released vapor, and controlling ignition sources. The likelihood of releases can be reduced by the methods described in these guidelines. Where feasible, substitution of process materials with nonflammable or less flammable substances makes the process inherently safer.

1.6.3 Explosions

Delayed ignition of a continuous or instantaneous release may result in an explosion (*vapor cloud explosion*, sometimes termed an unconfined vapor cloud explosion or UVCE). Whether or not an actual explosion with blast overpressures occurs appears to depend on the mass of flammable vapor in the cloud, the extent of dilution, the degree of turbulence and confinement, and possibly the nature and strength of the ignition source. This subject is dealt with more exhaustively in other texts (Gugan, 1979).

Historically, there appears to be a lower limit to the vapor quantities necessary to result in a vapor cloud explosion, although there is no established theoretical basis for this. For light hydrocarbons such as propane, about 5 tons can be inferred as a minimum amount of flammable material in a cloud to allow the flame front to accelerate to speeds required to generate blast overpressures. For ethylene, the minimum cloud size may be as small as one ton (Slater, 1978).

Whether or not confinement and/or turbulence is essential to the development of a vapor cloud explosion is still open to debate. However, experimental work indicates that partial confinement and obstacles do increase the flame speed in the cloud (Van den Berg, 1985). Flame speed is an important variable in determining the speed of the combustion process, which must be fast enough to overcome pressure relief by cloud expansion in order for blast overpressures to be developed. Pressure relief is also decreased by the presence of partial confinement, such as may be provided by on-site walls and structures.

The same mitigation measures for fires also apply to vapor cloud explosions. It should be noted that the explosion center may be a considerable distance from the release location if the vapors travel downwind before igniting.

1.7 ANALYSIS OF THE NEED FOR MITIGATION

A variety of measures for mitigating the risk of vapor cloud releases are described in the following chapters of this document. Before employing any of those approaches, however, it is important to know where they are needed most and which measures will be most effective for a particular need. To determine this, it is necessary to understand the ways in which a vapor cloud release can be initiated and how great a risk each potential release presents to the employees, public, environment, and facility. One tool available for this task is risk assessment, which provides a methodology for identifying the most important sources of risk. With this knowledge, it is possible to allocate a company's resources where they will do the most good.

Assessment methods have been developed for analyzing facility risks by identifying the type of accidents that can occur (for example, vapor cloud release), the probability that a particular accident will occur, and the accident consequences. These methods provide a systematic, thorough, and objective way to quantify risk and, in doing so, to indicate where mitigation measures are needed most.

Some of these risk analysis methods are described in detail in the AIChE-CCPS document *Guidelines for Hazard Evaluation Procedures* (AIChE-CCPS, 1985), and others will be described in a forthcoming AIChE-CCPS document entitled *Guidelines for Chemical Process Quantitative Risk Assessment Procedures*. A suggested risk analysis sequence and several risk analysis methods are discussed briefly in chapter 9.

1.8 VAPOR RELEASE MITIGATION APPROACHES

There are many ways to prevent or mitigate a release of hazardous materials. Release prevention (integrity) is a very important aspect of vapor release mitigation (Kletz, 1975). Methods for release prevention are emphasized in the AIChE-CCPS document *Guidelines for Safe Storage and Handling of High Toxic Hazard Materials*.

A distinction can be made between plant integrity and process integrity. The former implies mechanical integrity of the physical facility required to contain hazardous materials. The latter implies proper control of the process chemistry and thermodynamics (Lees, 1980, pp. 72, 76, 1048).

Both active and passive systems can be used to reduce the risk of accidental vapor cloud release. A useful distinction can be made between mitigation systems that require the proper functioning of personnel and/or equipment for adequate performance (active) and those which do not require manual actuation or utilities (passive). Where equivalent mitigation can be obtained by either an active or passive system, the passive system should be selected because it is typically more reliable. Passive systems may be more expensive than active systems; the cost of the protection is a factor in the selection process.

As a simple example, a method for reducing the rate of vaporization from a liquid spill is to limit the area occupied by the spill. A passive system for accomplishing this objective could be to install a dike or curbs around the potential spill source; an active method could be to provide sandbags near the potential spill source area, with the bags to be placed around any spill by personnel when they are needed. The active system in this example requires both time for action and people to be properly protected from the spill vapors.

Mitigation of releases can be achieved by actions taken in the design stage (prior to an incident) and/or by emergency-response actions (during and after the incident). Proper plant design can reduce the probability of a release and reduce the duration or rate of release. With prompt detection of releases, mitigation efforts can be initiated soon after a release has begun. Devices such as additional valves, containers, detection equipment, control systems, scrubbers, flares, and stacks can be incorporated at the design stage. Also, existing facilities can lessen the severity of vapor releases. Water and foam can be used to counteract releases, and existing valves can be used to shut off or divert the flow of fluids.

There is a preference order or hierarchy of approaches to mitigation of vapor releases. The chapters in these guidelines follow this general order.

- *Inherent Safety.* Eliminating the hazard by reducing the inventory of hazardous materials to nonhazardous levels would provide an inherently safe facility. Where this is possible, it is usually the most preferred approach. Some ways of accomplishing this are by minimizing vessel and piping volumes and by substituting nonhazardous materials. Siting the plant away from populated areas is another approach to inherent safety with respect to off-site risks. Mitigation through inherently safer plants is addressed in chapter 2.
- *Engineering Design.* Next in order of preference are the engineering design approaches which provide plant integrity and process (chemistry) integrity. These approaches will minimize the possibility of releases and provide engineered plant and process features that will minimize the consequences if a re-

lease does occur. These mitigation approaches are described in chapter 3.

- *Process Operation.* Improved process operation offers additional opportunities for preventing vapor cloud releases. Operating procedures, training, equipment testing and maintenance, and control of modifications are some of the essential administrative controls and management systems for risk management. Chapter 4 is a summary of process operating approaches to mitigation.
- *Emergency Action.* Prompt warning, activation of countermeasures, and contingency planning for emergencies are all part of a comprehensive approach to mitigation. Detection and warning systems are described in chapter 5, and guidelines for various release countermeasures are presented in chapter 6. On-site emergency response and alerting the surrounding community are addressed in chapters 7 and 8, respectively.

It should be noted that some mitigation features and approaches have been put in one of the above categories but may have fit just as well in other categories.

REFERENCES

AIChE-CCPS, 1985: *Guidelines for Hazard Evaluation Procedures*, prepared by Battelle Columbus Division for American Institute of Chemical Engineers--Center for Chemical Process Safety, New York.

Baer, B. M., and T. Kean, 1986: Toxic Catastrophe Prevention Act, New Jersey State Assembly, 13:1K-19.

CERCLA, 1986: Emergency Planning and Community Right-to-Know Act of 1986, Title III of the Comprehensive Environmental Response, Compensation, and Liability Act (CERCLA), Section 313, List of Toxic Chemicals, Washington, D.C.

Council of European Economic Communities, 1982: Council directive of 24 June 1982 on the major-accident hazards of certain industrial activities. *Official Journal of the EEC*, 82/501/EEC, No. L 230/1.

Fauske, H. K., 1985: Flashing flows or: some practical guidelines for emergency releases. *Plant/Oper. Prog. 4* (3), 132.

Gugan, K., 1979: *Unconfined Vapor Cloud Explosions*, Inst. Chem. Eng., George Godwin Ltd., Reading, U.K.

IChemE, 1986: Refinement of Estimates of the Consequences of Heavy Toxic Vapour Releases, Inst. Chem. Eng., North West. Branch, Jan. 8.

Kletz, T. A., 1975: Emergency isolation valves for chemical plants. *Chem. Eng. Prog. 71* (9), 73.

Lees, F. P., 1980: *Loss Prevention in the Process Industries*, Butterworth, London.

Slater, D. H., 1978: Vapor clouds. *Chemistry and Industry*, May.

United Kingdom, 1975: Health and Safety at Work, etc., Act of 1985.

Van den Berg, A. C., 1985: The multi-energy method--A framework for vapour cloud explosion blast prediction. *J. Hazard. Mater. 12*, 1.

Waxman, H. A., T. E. Wirth, and J. J. Florio, 1985: Toxic Release Control Act of 1985, U.S. House of Representatives Subcommittee on Health and the Environment, 99th Congress, H.R. 2576.

2

MITIGATION THROUGH INHERENTLY SAFER PLANTS

Eliminating a hazard altogether is obviously a desirable approach where vapor release hazards exist. In some processes, this is feasible by making significant design changes or using alternative process materials. Even if the hazard may not be entirely eliminated, methods such as inventory limitation and process modifications can make a facility inherently safer (Kletz, 1984).

2.1 INVENTORY REDUCTION

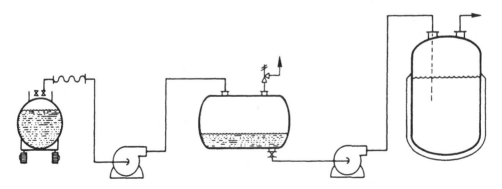

Limiting the inventory of hazardous chemicals may reduce the consequences of loss of containment. There are many ways and many places to accomplish inventory reduction; for instance, eliminating intermediate storage or generating hazardous material on-site only as it is needed.

A useful screening method for determining safe inventories of hazardous chemicals can be determined by using appropriate release and dispersion models to calculate the maximum quantity which could be released under "worst case" atmospheric conditions without exposing any population of concern to a dangerous concentration of the chemical.

A commonly used definition of a dangerous concentration is that which is considered "Immediately Dangerous to Life and Health" (IDLH). The National Institute for Occupational Safety and Health (NIOSH) has established IDLH values for many toxic materials (NIOSH, 1985). The IDLH is defined by NIOSH as the highest concentration to which a healthy worker can be exposed for 30 minutes and still escape without any permanent health effects. Because the IDLHs were established for the purposes of evaluating respiratory protection for normal healthy individuals, they are not directly applicable for hazard screening levels for off-site effects. New consensus planning guidelines are being developed by the American Industrial Hygiene Association (AIHA) to have more accurate concentrations for screening and planning purposes in situations involving on-site and off-site populations. These consensus values, known as Emergency Response Planning Guidelines (ERPGs), are expected to be made available on a continuing basis over the next few years. Each plant/company must decide on screening and emergency planning levels to be used at their facilities, with input from local toxicologists and other health professionals, or by acceptance of published guidelines.

Failure of existing facilities to pass such a screening test implies a possible need for secondary containment, isolation devices, transfer capability, countermeasures, and/or further assessment of risk (frequency of releases and magnitude of consequences).

A similar screening method could be used for flammable vapor clouds, using the lower flammable limit of the vapor as the screening criterion. (Note that the averaging time for dispersion modeling when dealing with flammable vapor clouds should approach zero, because momentary peaks above the lower flammable limit are sufficient to allow vapor ignition.) Limiting the maximum inventory so that release of the flammable material will result in an off-site concentration below the lower flammable limit implies that a flammable vapor cloud could not be ignited off site. Even so, off-site effects from an on-site flash fire or fireball, or blast effects from a vapor cloud explosion, may need to be taken into account.

2.2 CHEMICAL SUBSTITUTION

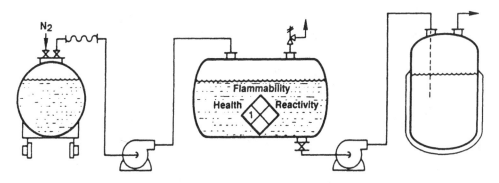

If less hazardous chemicals can be substituted in a process (Lees, 1980, p. 238), the consequence of an accidental release can usually be substantially reduced. The possible "side effects" of the change should be examined to assure that the change does, in fact, reduce the overall hazards of toxicity, explosion, and/or fire.

The following are examples where substitutions of materials have been employed to reduce the hazards of accidental release:

- Sodium hypochlorite ($NaClO*2.5H_2O$; solid oxidizer) is being used in place of chlorine (Cl2; toxic gas) for water treatment.
- Liquid urea (NH_2CONH_2; nontoxic solid) is being used in place of ammonia (NH_3; toxic gas).

2.3 PROCESS MODIFICATION

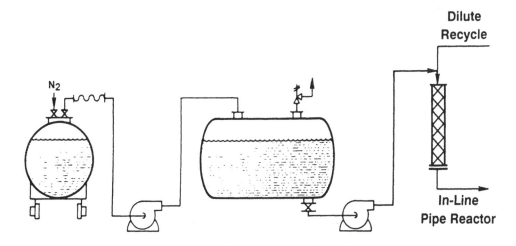

Modifying processes also can reduce the hazard of accidental release. For example, processes have been developed in which liquefied gases (such as chlorine) are refrigerated rather than stored in high-pressure tanks at ambient temperatures (Kletz, 1984; Woods, 1980) or are transferred as a vapor rather than as a pressurized liquid (Kletz, 1984). Also, processes have been developed which react highly toxic intermediates as soon as they are produced, instead of providing hold-up storage. Examples of this type of process substitution include the replacement of stored, liquefied phosgene (Kletz, 1984) or methyl iso-cyanate with gas in piping. Nitroglycerine is produced today by em-ploying a continuous operation with drastically reduced quantities of explosive in the process, rather than by an earlier batch process which was extremely hazardous. The limiting reactant, glycerine, is drawn into the reaction point by aspiration, such that the reaction is immedi-ately stopped if, for example, utilities are lost.

Processes can also be modified by altering the design conditions. Reducing the temperature in a process which handles a material at elevated temperatures, such as lowering it from above the boiling point of the process material to below the boiling point, will signifi-cantly reduce the amount of vaporization upon loss of containment. Likewise, bringing the operating pressure closer to atmospheric re-sults in a lower driving force and a correspondingly lower leak rate for a given hole size. Reducing extreme operating conditions or corrosiv-ity of process materials also generally results in a lower likelihood of release caused by deviations in process conditions beyond the design limits or deterioration of the process containment.

More detailed discussions of two process modification ap-proaches, refrigeration and dilution, are given in Sections 2.3.1 and 2.3.2.

2.3.1 Refrigerated Storage

Refrigerated storage is a means of reducing the rate of vapor release upon loss of containment of volatile materials. In this context, refrig-eration refers to the cooling of the process material to a temperature at or below its atmospheric-pressure boiling point. (The economic trade-offs between storing a material as a liquid rather than as a vapor are not addressed here.) Refrigeration has several possible benefits when applied to an installation handling liquefied gases:

- the driving force for leakage through a breach in the process containment is reduced at lower temperatures (and corre-spondingly lower vapor pressures), resulting in a lower leak rate;

- the amount of material which immediately vaporizes by adiabatic flashing upon release to the atmosphere is reduced or eliminated;
- the amount vaporizing as a mist is reduced by lowering the jet entrainment/mist generation which ensues when a pressurized release occurs;
- the initial rate of pool boiling and/or pool evaporation is significantly reduced; and
- the lower evaporation rate also allows more time for other mitigating measures, such as foam coverage of the spill, to reduce the total quantity of material released as a vapor.

Although refrigeration will, for most materials, reduce the vapor release rate upon loss of containment, it often necessitates insulation of process equipment and piping. Insulated equipment is more difficult to inspect, and embrittlement and external corrosion problems are more likely to be encountered; consequently, careful design is required for such installations. Further considerations related to refrigeration are as follows: the vapors released from a refrigerated process may be colder, and thus more dense, than a nonrefrigerated process; the cooling may possibly mask an incipient exothermic reaction; and the refrigeration equipment must be considered as to its reliability and the possible effects if loss of cooling was to occur. Nevertheless, when the proper design and materials of construction are used, refrigeration can be very effective. Historical leak experience with refrigerated ammonia storage, for example, has been very good at most installations.

2.3.2 Dilution

Dilution is another means of reducing the rate of vapor release upon loss of containment of volatile materials. In this context, dilution refers to the dissolving or mixing of the process material in another less hazardous material. Examples of dilution include

- handling ammonia as an aqueous solution (ammonium hydroxide) instead of as anhydrous ammonia,
- using various acids such as HNO_3 and HF in lower-than-fuming strengths,
- transporting highly toxic arsine gas in a diluent gas, and
- processing a flammable vapor at a concentration below its lower flammable limit.

For materials soluble in liquid solutions, diluting the toxic and/or flammable material generally reduces the partial pressure of the material above the solution. Consequently, if the solution is released from the process or storage containment, the vapor release rate will be lower for the lower-partial-pressure situation, since pool evaporation is a function of partial pressure and other factors such as wind speed and pool size (Opschoor, 1980). Also, dissolving a material in, for example, an aqueous solution may effectively increase the boiling point of the material, which further reduces the likelihood of adiabatic flashing and pool boiling. For example, anhydrous ammonia (NH_3) has a vapor pressure of *4600 mm Hg* at 10°C. Handling NH_3 as 60% ammonium hydroxide reduces the NH_3 partial pressure to *2400 mm Hg* at 10°C, and diluting to a 16% strength further reduces the NH_3 partial pressure to *90 mm Hg* at the same temperature.

Dilution can also be applied as a mitigation measure after a release has occurred. This is generally done by having, for example, fire-fighting water or dedicated fire-fighting monitors at a diked area to knock down vapors and dilute the spilled material. Experimental work may need to be performed to determine the concentration at which significant vaporization is suppressed.

Heat is liberated when diluting most common acids and other water-soluble flammable/toxic materials. For this reason, the temperature increase must be considered when evaluating dilution as a mitigating measure (when performing vaporization calculations).

2.4 SITING CONSIDERATIONS

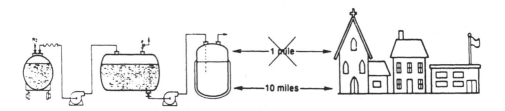

Isolation (by distance) of a chemical process from on-site and off-site surrounding populations is generally a very effective consequence-mitigation measure. Separating the process from vulnerable populations affords both attenuation of the effects and time to provide emergency response. The reductions in the blast effects of an explosion, the thermal radiation from fire exposure, and the vapor concentration of a toxic release are more than proportional to the distance from the hazard source.

Providing a buffer zone between a process and *on-site* populations is likely to be effective only for equipment-explosion hazards and thermal radiation from fire (Lees, 1980, p. 211). For adequate isolation from toxic vapor clouds or vapor cloud explosions, distances of several hundred feet or more may be required for a significant reduction in toxic vapor concentrations or blast pressures. That is, the distance of 100 feet suggested for separation of processes for prevention of accident propagation by thermal radiation (Lees, 1980, p. 217) is unlikely to provide a significant reduction in hazard or a significant amount of warning time for self-protection of potentially exposed personnel (100 feet is equivalent to less than 15 seconds at a typical wind speed of 5 miles per hour). The measures that can be taken and the procedures that should be developed and tested are discussed in chapter 7.

Thus, isolation by distance is not likely to be a practical method of vapor cloud mitigation for on-site exposure, so other hazard-mitigation measures should be provided. For example, explosion hazards may warrant the use of blast-resistant buildings. Toxic release hazards may require providing safe havens and escape respirators (chapter 7). For off-site exposure, attempts should be made to secure as much isolating distance as practical, based on inventory, average wind speed, toxicity, and land availability and cost, particularly in the prevailing downwind direction. Some process companies use hazardous event consequence calculations to aid in planning the layout of their facilities. For example, a hazards analysis early in the design stage may identify one particular unit as having the greatest potential for a toxic release, and that unit may then be located on the site as far as possible from off-site neighbors, perhaps considering prevailing wind directions as well. Similar considerations may be included when laying out relative locations of on-site facilities. This is a common practice in the explosives manufacturing industry, which uses standard minimum distances for separation of different building types.

REFERENCES

Kletz, T. A., 1984: *Cheaper Safer Plant or Wealth and Safety at Work*. Inst. Chem. Eng., London, 1, 6, 19, 53, 54, and 69.

Lees, F. P., 1980: *Loss Prevention in the Process Industries*, Butterworth, London.

NIOSH, 1985: *Pocket Guide to Chemical Hazards*, DHHS Publ. No. 85-114, U. S. Department of Health and Human Services-National Institute for Occupational Safety and Health, Washington, D. C.

Opschoor, G., 1980: Evaporation. Ch. 5 in *Methods for the Calculation of the Physical Effects of the Escape of Dangerous Material* (Liquids and Gases), Netherlands Organization for Applied Research (TNO), Voorburg, The Netherlands.

Woods, B., 1980: Atmospheric storage. *Proc. 23rd Chlorine Plant Oper. Semin.*

3

ENGINEERING DESIGN APPROACHES TO MITIGATION

If it is not possible to eliminate or sufficiently reduce the hazards by making the plant or process inherently safer by the methods described in the previous chapter, there are a variety of engineering approaches to mitigation that can be followed. While the facility is still in the design stage, plant integrity can be addressed to assure that the probability of a release is minimized. Attention to design and construction codes is an important aspect of this. Materials of construction can be specified which will maintain plant integrity while containing the process materials not only under normal process conditions but also during process upsets.

Process integrity can also be addressed in engineering design. Process integrity involves the chemistry of plant design and operation. Assuring that only the proper reactants and solvents are used and that they are of the required purity for the process and equipment, knowing and maintaining the conditions of operation within limits that are known to be safe, and avoiding processes that are sensitive to parameter deviations and providing measurement and control of those parameters are all important in avoiding upsets that lead to loss-of-containment accidents. Pressure relief systems, properly chosen and configured and venting into other containment or disposal systems, such as scrubbers, stacks and flares, are another way of enhancing process integrity.

Mitigation after loss of containment can also be effective and usually must be provided for in the process design stage. Secondary containment by double-walled (concentric) piping or double-walled vessels may be warranted. Dikes, curbs, and trenches leading away from storage vessels to strategically located impoundments can be used to reduce the rate of evaporation, help keep the liquid source of the vapor away from the most sensitive areas of the plant, and limit the extent of necessary emergency response activities.

These and other mitigation measures which can be effective if considered and incorporated into the engineering design are described in this chapter.

3.1 PLANT PHYSICAL INTEGRITY

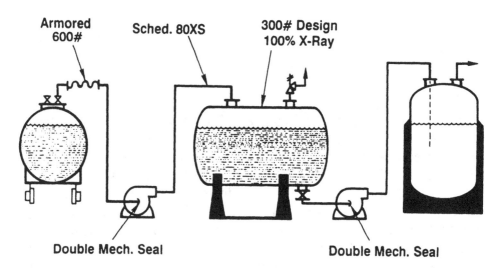

Since loss of containment is the chief cause of hazardous vapor cloud generation, maintaining the integrity of process equipment handling volatile material is obviously of fundamental importance in minimizing the likelihood of a vapor release. Various loss-of-containment causes are outlined in Appendix A. It is not the purpose of these guidelines to address the many details of maintaining plant integrity; this subject is dealt with in another AIChE-CCPS volume, *Guidelines for Safe Storage and Handling of High Toxic Hazard Materials*. This section covers only a few of the more important considerations.

3.1.1 Design Practices

Plants anticipating a high potential risk associated with release of process material may need to employ design features which are above and beyond standard practice for plant integrity. New equipment and methods are being developed continually, such as "high-integrity" flange systems and "fire-safe" valves.

Double Containment. If the consequences of process-equipment failure are exceptionally severe, because of large quantities of flammable or toxic materials or a high degree of toxicity or where even a small leak could have serious consequences such as inside an occupied building, double containment may be warranted. For example, large refrigerated ammonia tanks and cryogenic liquefied natural gas tanks are sometimes enclosed in secondary containers, with methods for detecting leaks in the annular spaces (Lees, 1980, p. 774; Trbojevic and Maini, 1985).

Double-Walled Piping. Greater-than-normal protection against piping failure is possible with double-walled piping, with the process fluid in the inner pipe and devices (such as analyzers and pressure switches) connected to the annular space to detect leakage. This is used frequently for underground piping and, in a few cases, for toxic gases in congested or enclosed areas with aboveground piping. When double-walled piping is used, provisions should be made for detecting, locating, and isolating leaks.

Seismic Effects, Wind Loading, Subsidence. The ability of process equipment to withstand earthquakes and severe winds is best considered in the design stage. The design criteria will, of course, be dependent on the location of the facility. The design of all equipment handling material with significant hazard potential should also include consideration of subsidence (settling or tilting). This particularly applies to equipment on filled ground or where a high water table requires extensive use of piles for equipment support. Subsidence checks for storage tanks, stacks, and columns should be included in a plant's periodic inspection program.

3.1.2 Materials of Construction

The use of improper materials of construction can lead to premature and possibly catastrophic failure of piping or process vessels. Causes of sudden failure include handling corrosive materials at high temperatures and pressures, processing materials that contain corrosion catalysts such as chloride ions or active hydrogen, and residual stresses in equipment (Perry and Green, 1984, pp. 6-96, 6-109, 23-4, 5, 7, and 12). Brittle fracture can occur from exposure to low temperatures (cryogens), caustic or nitrate solutions, hydrogen, fatigue, or vibration. Biological attack (from organisms in oxygen-rich organic/aqueous mixtures) also can lead to equipment deterioration.

Selecting the best materials for a given application is the primary method of protecting against corrosion and erosion. Corrosion testing is often a crucial part of selecting appropriate materials. Wherever possible, a sample from the actual process stream should be used in the corrosion tests. Considerable data are available on materials of construction for piping and process equipment that are resistant to the wide variety of corrosive materials encountered in chemical plants (Baumeister, 1979; Gackenbach, 1960; Hamner, 1974; Hamner, 1975). For many combinations of ferrous and nonferrous metals and process liquids, the effects of concentration and temperature on the rate of corrosion are well known (Perry and Green, 1984, pp. 6-96, 6-109, 23-4, 5, 7, and 12).

Plants anticipating a high potential risk associated with release of process material may need to specify materials of construction which give a greater degree of protection against corrosion, embrittlement, etc., than found in plants handling less hazardous materials. This may involve more "exotic" materials of construction not only for piping and vessels but also for seals, gaskets, bolts, vessel internals, etc. A potential pitfall in using more exotic materials is that they are usually more difficult to fabricate and repair correctly, thus making, for example, weld failures potentially more likely.

Another consideration is to specify materials of construction which will ensure "leak before break," that is, the material will remain ductile under the lowest anticipated ambient and process temperatures, including cooling effects caused by expansion and evaporation of leaking fluids. Thus, failure of the containment at one location will remain localized rather than result in brittle fracture and total, sudden release of the process material.

Care must be exercised in choosing the insulation material for a particular installation, to ensure that it will not react adversely with the process fluid should the latter leak from the process. For example, some commonly used insulations can lower the apparent autoignition temperatures of some kinds of chemicals by promoting partial oxidation of hydrocarbons to aldehydes.

Inspection and Testing during Construction. Checks must be made to ensure that the correct type and quality of construction materials are used when building or modifying a facility handling hazardous materials. This includes quality and material assurance efforts during purchasing, fabrication, and installation, and covers not only piping and vessels but also welding rods, bolts, valves and internals, etc. Prestart-up verification, for example by X-ray and dye penetration tests, are often required to conform to a given code or attain a design rating or service category.

3.2 PROCESS INTEGRITY

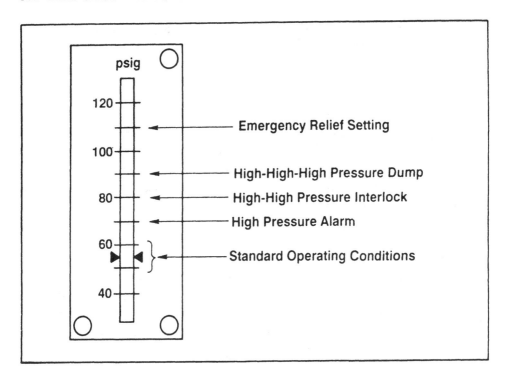

Process integrity refers to controlling the chemical (as opposed to the purely mechanical) aspects of a process to maintain containment of the process material. Aspects of process integrity include controlling the identity and purity of reactants and solvents, establishing the correct envelope of operating conditions within which the process can be safely operated, providing process control systems and operating procedures to keep the process within the operating limits, and using emergency controls such as interlock and overpressure protection systems to ensure that the design limits of the containment are not exceeded for normal as well as for emergency operations.

3.2.1 Identification of Reactants and Solvents

Use of a wrong or insufficiently pure substance in a chemical process could result in undesired reactions, excessive corrosion, or other possible causes of a vapor release. A good practice to prevent misloading from a supplier's container (tank car, tank truck, barge, cylinders, etc.) into the wrong tankage or process is the use of special fittings, well-marked connections and valving, and the careful examination of shipping papers and containers. Additional on-the-spot analyses (such as checks for density, color, pH, pressure, etc.) may be warranted if the consequences of process contamination or incompatibility could be serious. Also, many plants require that unloading connections be made by operating personnel or, at least, that an operator be present when the unloading connection is made.

3.2.2 Limits on Operating Conditions

In a well-designed process, the designer will have established an "envelope" of operating temperatures, pressures, flows, and other parameters that will provide both safe operation and efficient production. Additional testing may be warranted if information concerning heat of reaction, autoinitiation (or exotherm onset) temperature, or "temperature of no return" are not known for a process which might have runaway or explosion potential.

Controls and procedures are then provided to assure that the process is operated within these limits, with emergency controls in place to shut down the process and/or protect the process equipment and personnel from misoperation. Where possible, designing the process such that potential deviations from normal operation will be physically limited provides an inherently safer design. Some examples which have been successfully used (depending on the specific application) are

- using hot water instead of steam to heat a process material, in order to limit the maximum temperature;
- where steam is used as a heating medium, using a lower pressure steam supply (desuperheating may be required for large reductions in pressure);
- installing a relief valve on a padding gas supply line to prevent vessel overpressuring or vessel relief device actuation; and
- specifying smaller line sizes, restricting orifices, or metering pumps to limit the release rate if a line break occurs.

For autoreactive or sensitive materials (such as monomers, some catalysts, and potentially explosive materials) and strongly exothermic reactions, appropriate operating limits, particularly temperature, often can be established from the results of small-scale tests. For example, the accelerating-rate calorimeter (Fenlon, 1984) provides an adiabatic environment (simulating a large container) for evaluating self-heating while measuring temperature and pressure. Other methods for evaluating thermal hazards include Constant Temperature Stability (CTS), Differential Thermal Analysis (DTA), heat accumulation, classical chemical analysis (gas and liquid chromatography), and isothermal aging techniques (Hub and Jones, 1986). On-line methods for monitoring exothermic reactions have also been used.

3.2.3 Process Control Systems

Effective process control systems can reduce the frequency of chemical releases by maintaining the process within established safe operating limits. Failure of process controls can lead to such events as uncontrolled exothermic reactions, vessel overflows, and releases from relief devices such as relief valves, rupture disks, and explosion vents.

The frequency of process-caused releases usually can be reduced by careful specification of additional process controls and interlock shutdown devices (Kohan, 1984). Either the same type of sensor (redundancy) or a sensor operating on a different characteristic of the process (diversity) can be used. An example of the latter would be an inferred "high-temperature" interlock based on excessive vapor pressure (Prugh, 1974). However, care needs to be exercised that increased complexity does not somehow increase rather than reduce the overall risk. Elaborate systems are sometimes very difficult to maintain. All process controls need to be functionally tested on a periodic basis to maintain their reliability; this testing must be considered in the design of the controls. Operator response effectiveness is decreased when too much process information is presented without an alarm hierarchy or other aid. For these reasons, a simpler, inherently safer process modification is usually preferable over increased complexity of process controls.

Computers are increasingly being used to monitor unsafe trends in process variables and thus help avoid accidental releases. However, the use of computers with process safety systems is a subject which needs very careful analysis. Some operating companies have chosen to rely solely on "hard-wired" devices, such as electromechanical relays and pressure switches, for critical safety interlocks, because of possible undetectable faults or common mode failures which are conceivable with systems which are entirely computerized.

Control systems in many of today's integrated process plants are very complex, and thus it is difficult to check the adequacy of a design to ensure process integrity. Systematic logic modeling and risk assessment tools such as fault-tree analysis are especially well suited to study complex control systems and find weaknesses in their reliability.

3.2.4 Pressure Relief Systems

3.2.4.1 RELIEF DEVICES

Pressure relief devices have a long history of providing protection from overpressure. Relief valves are commonly used for overpressure protection because they are self-closing, and thus can limit the duration of the release. If fast action or considerable capacity is required (e.g., to relieve a runaway reaction or a vapor-space deflagration), rupture disks may be the necessary choice. In the past, such devices were relieved directly to the atmosphere. More recently, however, installations with potential for release of hazardous vapor are being studied for any possible external effects, even though releases may be very infrequent.

If leakage from a relief valve cannot be tolerated, or if it is desired to isolate the relief valve from the process to prevent corrosion, a rupture disk may be installed between the process and the relief valve. However, careful attention must be paid to the design and maintenance of such an installation to prevent disabling of the relief protection. If the pressure in the chamber between the rupture disk and the relief valve increases above atmospheric pressure (e.g., because of a pinhole in the rupture disk), the effective burst pressure of the rupture disk will increase accordingly (Lees, 1980, pp. 312, 616) because the rupture disk bursts only when the *differential* pressure between the upstream and downstream sides attains the disk rating. Thus, it may be possible to exceed the rupture pressure of the vessel before the rupture disk bursts. Either this chamber must be vented (to the atmosphere or to a treatment system reliably maintained at atmospheric pressure) or the pressure in this chamber must be alarmed and/or monitored.

Specific installations may warrant special designs to ensure the reliability of the relief device operation and to prevent secondary effects following actuation of the device. Some types of rupture disk (e.g., graphite, ceramic, and some styles of frangible disks) may yield fragments that can clog or jam the relief valve; in these cases, a fragmentation chamber can be designed to protect the relief valve by trapping the fragments. However, it is preferable to specify a nonfragmenting disk. If the relieving material can be self-ignited or react with air following the opening of a rupture disk under a relief valve, the chamber between the disk and the valve should be evacuated and charged with a nonreactive, compatible gas. An example of this situation is the decomposition of tetrafluoroethylene, which has been known to initiate by adiabatic compression upon bursting of a rupture disk with air present in the chamber.

Careful attention must be given to the selection and sizing of emergency relief devices, particularly where two-phase flow is possible. Under the auspices of the AIChE, the Design Institute for Emergency Relief Systems (DIERS) has completed a significant amount of research into many aspects of relief sizing (Fisher, 1985).

Reliable operation of an emergency relief system depends not only on relief device(s) which are properly selected and correctly installed, but also on the prevention of upstream and downstream blockage due to line plugging or a closed valve. The use of valves in lines upstream or downstream of a relief device must be carefully monitored and controlled; this subject is discussed in the AIChE-CCPS volume titled *Guidelines for Safe Storage and Handling of High Toxic Hazard Materials*.

3.2.4.2 RELIEF HEADERS

Currently, direct discharge to the atmosphere from relief devices, emergency depressuring systems, etc. (referred to as *discharge systems* in the following) is not desirable in many cases. Economics dictate the use of common treatment/disposal facilities, and this, in turn, requires the use of relief headers (collection systems).

The design of header systems requires consideration of many factors:

- Each system must have the capacity to handle the anticipated instantaneous load. In the worst case this is the actual capacity (which may be significantly greater than the design capacity) of every connected discharge system. It may be possible to reduce the anticipated load by providing interlocks that reduce the probability of simultaneous releases to an acceptable level (this may require a quantitative risk assessment), by performing analyses that show times-to-release of various operations will result in acceptable staggering of the releases, etc.
- Imposing a significant back pressure on a rupture disk is usually unacceptable, because of its effect on the bursting pressure. Therefore, it may be necessary to have a separate line from each rupture disk to a point of assured near-to-atmosphere pressure (e.g., the inlet to a flare). Once the disk has burst, the only effect of back pressure is to reduce the flow through the disk. This permits optimizing the size of the rupture disk and its discharge line (which are generally the same) to achieve the desired flow rate.
- Imposing a back pressure on a relief valve can cause a change in the opening pressure and can cause the valve to close prematurely. The American Society of Mechanical Engineers (ASME) Code limits the back pressure on a relief valve to 10 percent of the set pressure, unless the valve has a pressure-

compensating bellows or similar protection against the effects of back pressure (ASME, 1974). Pilot-operated relief valves may easily be designed to be immune to back pressure, but they have other limitations (e.g., clogging of pressure-sensing lines) that may prevent their use in some services. Relief valve capacity formulas assume that sonic flow occurs in the nozzle of the valve, unless correction factors are applied; thus, if the back pressure (absolute) is greater than approximately one-half of the inlet pressure (absolute), the capacity of the relief valve may be reduced.

The foregoing provides many opportunities for optimization of header systems in complex plants. For example, two or more systems may be provided to service valves in high pressure and low pressure operations; or, one or more low pressure valves may be oversized to allow subsonic flow in their nozzles (however, the possibility of much larger flow under low back pressure conditions must be considered).

Capacity is not the only consideration in designing header systems:

- Some discharges may be cold enough to require the system to be made of low-temperature alloys, in order to avoid embrittlement (e.g., depressurization of liquid ethylene, even if the liquid is above ambient temperature). Economics may dictate two systems, one for cold and one for ordinary discharges (Feldman and Grossel, 1968).
- Discharges containing water (liquid or vapor) or other materials subject to freezing should not be mingled with cold discharges.
- Freeze protection such as electrical tracing may be needed in some installations where leakage into the relief header may cause plugging of the header by solidified process material.
- Corrosive discharges may justify separate systems.
- Incompatible materials should be segregated. This applies to acids/alkalis, oxygen/hydrocarbons, chlorine/aromatic hydrocarbons, etc.
- It may be desirable to segregate plugging materials (e.g., discharges from polymerization reactors) in order to minimize the cleaning problem or the need to shut down an entire unit or plant to clean the header system.
- Unless the possibility of having liquid in the header system can be absolutely eliminated, it is advisable to have the discharge for each system run downhill from the device to the first part of the header/treatment/disposal system that is intended to han-

dle liquid. In addition to the possibility that pockets of liquid may affect the operation or capacity of a relief device, the acceleration of a slug of liquid may damage the piping.

- The potential for reverse flow into other vessels or into lines from other relief devices connected to the header requires careful analysis and may be another reason for specifying separate relief headers.
- Mechanical reaction forces in piping downstream of relief devices and vessel supports must be carefully analyzed. Failure of a relief header due to excessive reaction forces can result in direct discharge to the atmosphere in an unknown direction, and may have a domino effect (e.g., if a flammable vapor is continuously relieved and ignites following header failure). Stress analysis is often warranted in critical safety systems.

3.3 PROCESS DESIGN FEATURES FOR EMERGENCY CONTROL

If a process becomes "out of control" and exceeds its safe operating limits, various engineered features can be employed to mitigate the effects of an emergency situation that has the potential for vapor release. These include not only emergency shutdown and safe disposal of emergency releases, but also means to quickly stop or substantially reduce the release rate and/or quantity if loss of containment occurs.

3.3.1 Emergency Relief Treatment Systems

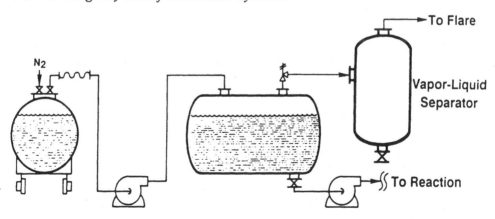

Relief headers (as discussed in Section 3.2.4.2) and discharge piping from individual relief devices may be routed to an emergency relief disposal system such as a scrubber, flare, or incinerator which is capable of eliminating or substantially reducing the toxicity and/or flammability hazards of the vapor released in an emergency situation.

The possibility of liquid entrainment or condensation in the discharge stream usually requires the use of knockout drums or catchtanks upstream of the disposal system. Although some of the process equipment discussed here (such as scrubbers and absorbers) are usually designed for normal process operation, they can also be designed to handle exceptional, short-term, high-throughput situations which are characteristic of emergency releases.

Another possible function of some disposal system components is to use the volume of the system as a "load-leveling" mechanism, allowing a short-duration, high-rate release to be distributed over a longer period of time, thus putting less of a shock and burden on the system. This function is possible not only with active or passive scrubbers, but also with other means of providing volume capacity including gas holders, larger diameter piping, and catchtanks. Such systems require careful design to be effective, and possible side effects such as increased gas inventory may need to be considered.

Combinations. Systems in which filters, coolers, condensers, receivers, scrubbers, and adsorbers perform separate functions in removing particulates, water vapor, high-boiling-point materials, water-soluble fluids, and other toxic or flammable materials may be more efficient and have greater flow-rate capacity in handling some types of emergencies (such as fires or runaway reactions in confining structures), as compared to single units. Careful study of the physical and chemical properties of the stream to be treated, along with testing of the system, is needed to determine the optimum arrangement (flow order) and equipment sizes.

3.3.1.1 ACTIVE SCRUBBERS

A scrubber provides good contact between a vapor or gas and a liquid in which the vapor or gas (or certain components) can dissolve. The scrubber may be a packed tower, a plate column, a spray chamber, a wetted-wall column, a stirred or sparged vessel (Perry and Green, 1984, pp. *3*-14, *14*-6, *14*-14, *14*-18, *14*-31--*14*-34, *14*-38, *18*-3, and *20*-89), or a venturi. Usually, the flows are countercurrent, with the gas entering at the bottom and the liquid entering from the top (Wilke and Von Stockar, 1985). Water is used for the absorbing liquid if the gases or vapors are soluble in water; caustic solutions (NaOH) are used to rapidly neutralize acid vapors. Sometimes organic liquids with low vapor pressure are used to absorb organic vapors. A packed tower should not be used where a discharge stream may contain solid particles (e.g., catalyst) due to possible plugging.

Design. Scrubbers are sometimes designed to receive gases and vapors from emergency systems, such as relief or pressure control valves. If there are multiple valves, the scrubber should be sized for the largest release. In any case, emergency-venting flows are likely to be much larger than typical maintenance-preparation flows, with correspondingly larger scrubbing equipment (towers and recirculating flows) required.

Operation and Safety Considerations. Emergency scrubbers can be designed to operate continuously or operate on demand (standby). Continuously operating emergency scrubbers are used in facilities protected by depressuring valves, relief valves, and/or rupture disks where sudden large flows may occur. Alarms should be installed to signal low circulation flow, abnormal scrubber temperature, and depletion of scrubbing solution. If the hazards of operating without a scrubber are severe, plant operation may need to be interrupted when the scrubber is not available.

Standby manual emergency scrubbers are used in "batch" type operations, such as unloading tank cars or preparing piping or equipment for maintenance. Typically, the scrubber may be started before each unloading or other operation begins. The system should have operating lights to signal proper operation and alarms to indicate scrubber failure. In the event of scrubber failure, the operations protected by the scrubber should be interrupted until the scrubber can be made to function normally.

Standby automatic emergency scrubbers are used to treat air exhausted from enclosures. Although exhaust blower operation would be continuous, the scrubber circulation would begin only in response to a signal from a vapor detector or other indication of leaking. Standby automatic scrubbers should be tested periodically to ensure that they respond appropriately to alarms and signals. Backup (duplicate) detectors should be installed to ensure triggering of the automatic standby system. Dual circulation pumps may also be warranted.

For all types of emergency scrubbers, the scrubber and piping should be designed to cope with emergency flows that occur at high rates. Slugs of liquid or two-phase flow discharging into scrubbers designed for vapor could cause severe temperature increases (localized boiling) with correspondingly poor absorption and breakthrough. Flooding may also occur at high vapor rates, carrying scrubber liquid into discharge stacks. An effective emergency scrubber must be designed to avoid these situations.

3.3.1.2 PASSIVE SCRUBBERS

In many cases where the disposal of a high rate of vapors is necessary, a scrubber as described in the preceding section will either be too large to be economical or practical, or will not by itself give sufficient fractional reduction in the amount of vapors released to the atmosphere. It also may be uneconomical or undesirable to have a continuously operating scrubber in a system where the scrubbing function is expected to be needed only rarely in emergency situations. In addition, a greater degree of reliability on demand may be needed in release scenarios involving, for example, plant electrical failures which may both cause an emergency venting and disable the scrubber circulating pumps. Passive scrubbers have been used by some chemical companies in these cases as an alternate means of scrubbing the vapors from an emergency release. The primary benefits of a passive scrubber are reliability (being a passive device) and reduced operating costs.

A passive scrubber, sometimes called a "scrub tank" or "guard tank," is a process unit intended to scrub a relatively large-quantity, short duration throughput of vapor without necessitating active components such as recirculation pumps. A passive scrubber might consist of no more than a tank containing a liquid scrubbing solution through which the vented vapors are sparged, with the vapor outlet going directly to the atmosphere through an elevated vent pipe. Passing the vented stream through the scrubbing solution may have one or more possible functions, including (1) reacting with an undesirable vapor component to detoxify and/or neutralize it, (2) absorbing the vapor into the scrubbing solution, (3) condensing the liquid component of a two-phase stream so only the vapors pass through the tank, and (4) killing a reaction which is proceeding in the vented stream. The scrubber may be designed for total vapor removal or for selective removal of one or more components of a multicomponent stream.

Discharge from a passive scrubber may be routed to an "active" scrubber or flare if further vapor removal is necessary. Further treatment is usually warranted when the passive scrubber is used for absorbtion or condensation.

Design. The description of a passive scrubber as a mere sparge tank can be misleading as far as the design considerations. The passive scrubber must be designed very carefully to obtain the desired degree of capacity, efficiency, and back pressure. Since the vapors only pass through the tank once, with no packing or other means of enhancing the degree of vapor/liquid interfacial surface area, extremely good phase contacting and mixing action is necessary. In particular, the sparger design is critical to obtain the proper jet mixing as a function of the size and number of holes and the back pressure. Liquid height (to determine contact time and back pressure), liquid volume (to determine total scrubbing capacity), and vessel diameter (to avoid entrainment) are also design variables.

The gas distribution is accomplished by an open-ended standpipe, a horizontal perforated pipe, a perforated plate at the bottom of the tank, or porous septa (Perry and Green, 1984, pp. *18*-61--*18*-63), or by a bubble-cap distributor. The sparger design must allow for drainage since it will be liquid-full when not in use.

Maximum scrubbing capacity is relatively easy to calculate if, for example, the neutralization reaction is known. However, no design data may be available on passive scrubber efficiency in a specific application, possibly necessitating test work to demonstrate a given efficiency, such as 90 percent removal of chlorine from a vent stream using a dilute caustic solution.

Operation and Safety Considerations. The scrubber acts in a batchwise mode; that is, since it is expected to be used very infrequently, it does not have a continuous make-up stream of fresh solution, but rather the entire contents would normally be replaced after an emergency venting occurs.

A definite advantage of a passive scrubber as part of an emergency relief system is that it requires no active components such as pumps or valves to function properly. However, it will have little or no benefit if the scrubbing liquid is not in the tank and is not "potent" at the time of a release. Consequently, administrative controls are necessary to ensure that a liquid level is maintained in the tank whenever the process is in operation. Possible problems include the following: the scrubbing liquid may be lost via leakage or evaporation, or may not be replaced after a shutdown or after maintenance on the vessel; any reactive components may become ineffective over time; or recharge of the scrubbing liquid may be neglected after a previous demand on the system. Hence, routine level checks must be made, and periodic checks of, for example, the caustic strength may be necessary. In safety-critical processes, an interlock may be warranted which would, for example, not allow start-up of the process without sufficient liquid level in the passive scrubber.

3.3.1.3 STACKS

Prior to the 1960s, stacks were widely used to disperse gases and particulate materials from power plants, gases and vapors from chemical plants, and particulate solids from a variety of dust-generating processes. At present, stacks are used to disperse the relatively low concentrations of gases, vapors, and dusts remaining after treatment by scrubbers, incinerators, precipitators, filters, and other devices. These types of treatment can remove 90 percent or more of the air pollutants. Occasionally, stacks are used for "last resort" emergency venting from chemical plants.

The following potential hazards, particularly if there is no treatment of the release prior to the stack, need to be evaluated before considering use of a stack (Grossel, 1986; API, 1982, p. 60; Wells, 1980, pp. 123 and 136; Lees, 1980, pp. 168, 315, 317, 530, 916, and 919; Bodurtha, 1980, p. 101; Jenkins et al., 1977):

- formation of flammable mixtures at ground level or at elevated structures,
- exposure of plant personnel and/or the surrounding population to toxic vapors or corrosive chemicals,
- accidental ignition of vapors at the point of emission (i.e., by lightning or static electricity), and
- air pollution at upper elevations.

Stacks have the following significant advantages over low-elevation releases such as releases at ground level:

- *Plume rise.* If the release has a significant velocity at the top of the stack, a plume rise is created that increases the effective height of the stack. Gases that are less dense than the atmosphere (e.g., because of high temperature or low molecular weight) create a draft by buoyancy that propels the discharge upward (Perry and Green, 1984, pp. 9-72, 26-20). If necessary, a blower or compressor may be used to obtain the desired discharge velocity.
- *Jet entrainment of air.* The velocity of the rising plume also entrains some of the surrounding air, resulting in a dilution effect (Hoot and Meroney, 1974). This benefit can be further enhanced by reducing the tip diameter (commensurate with back pressure considerations), thus increasing the gas exit velocity.
- *Dispersion.* Gas released at an elevation substantially above ground level greatly reduces the ground-level concentration of toxic gases (at the same downwind distance) due to the effect of vertical and lateral dispersion (Lees, 1980, pp. 220 and 432). However, stacks do not reduce the amount (rate or quantity) of pollutants released.
- *Upper-air wind speed.* The wind speed at high elevations usually is much greater than the speed near ground level; typically,

50 percent higher at 50 meters than at a reference height of 10 meters during the day, and as much as 8-fold greater at night (API, 1982, p. 61).

- *Wake effects.* Tall stacks can avoid the "ground-looping" effects of unfavorable topography or building configurations, provided that the top of the stack is well above the height of the nearby structures.

Stack Design. Before designing a stack, the acceptable ground-level concentration for the gases, vapors, or particulate material to be discharged must be determined (API, 1982, p. 61). For nontoxic flammable or combustible material, this concentration would be the lower flammable limit (Bodurtha, 1980, p. 98) divided by a safety factor which accounts for time variability of concentration in eddies. *Guidelines for Use of Vapor Cloud Dispersion Models* (AIChE-CCPS, 1987, p. 94) contains more information about the time variability of concentrations at any point. Concentrations of flammable materials which are higher than the lower flammable limit could lead to accidental ignition. The stack height calculated by this design procedure may have to be increased to protect personnel on the ground or on nearby elevated structures from thermal radiation.

Concentrations of toxic materials released only under emergency conditions should not exceed dangerous concentrations, as discussed in Section 2.1.

A stack should not be located downwind (based on the prevailing wind direction) of hills or tall buildings that might cause air turbulence (API, 1982, p. 61). Stacks that discharge corrosive materials should not be located upwind of cooling towers or air-cooled heat exchangers (Bodurtha, 1980, p. 98). If the stacks are located near or within the flight paths of local airports, they may require warning lights.

To prevent saturated vapor leaving the separator from condensing inside the stack wall and being blown out as an aerosol, some stacks need to be insulated and steam traced so that the temperature of effluent vapors is always above the dew point. A less expensive but also less effective alternative is to use a skimmer-type separator at the stack tip. Skimmer-type separators capture condensed vapors that creep up the stack wall and drain them externally. Further details concerning stack design are given in the references (API, 1982, p. 61; Slade, 1968; TNO, 1980).

Operation and Safety Considerations. Periodic inspection of stacks should be performed to assure adequate maintenance given the ambient environment and the materials handled in the stack. Environmental conditions may include lightning, corrosive atmosphere, and erosion by wind and rain. Materials handled by the stack may have corrosive, erosive, oxidizing or reducing, acidic or alkaline, drying or moisture-containing properties. Inspection of steel or metal stacks should include wall-thickness measurements. Plastic or fiberglass stack inspection should include checks for delamination, melting, and chemical attack. Masonry stacks should be inspected visually for damage from deterioration and lightning. All stacks should be periodically checked for verticality, and any guy wires tested for proper tension.

3.3.1.4 FLARES

The use of flares is an effective means of eliminating combustible materials. They can also be effective in treating the release of toxic materials, particularly if such materials can be detoxified by burning. A flare system consists of a stack, with pilot flames at the tip, and an inlet header. Flares are most often used to burn combustible materials vented during

- start-up,
- normal conditions (i.e., vapors from tanks being filled),
- shutdown, and
- emergency conditions such as tanks exposed to fire or overheated reactors.

The primary concerns relating to flare operations are

- exposing personnel to thermal radiation at ground level or on nearby elevated platforms,
- exposing personnel to toxic combustion products or to unburned toxic components of the flare-feed gases,
- noise from the flare,
- flashback into feed lines,
- flameout,
- smoke or soot, and
- hot or burning liquid discharges.

Visible light, particularly at night, also may be of concern in some locations, but would not be an important consideration during emergencies.

Flare Design. The use of flares as a method of destroying hazardous material from process equipment is generally limited to flammable or readily combustible fuel materials, such as hydrocarbons. However, flare disposal has also been extended to flammable toxic materials, such as hydrogen sulfide, ammonia, carbon monoxide, and nickel carbonyl. Even for combustible materials, however, complete conversion to carbon dioxide and water vapor is unlikely. Therefore, the flare stack should be built high enough to avoid hazardous or "nuisance" concentrations of noxious gases (burned or unburned) at ground level.

Flares can also be used for emergency dispersal and partial oxidation of relatively noncombustible materials (less than 200 Btu per cubic foot), such as phosgene, chloroform, and methylene chloride, if they are supported by auxiliary fuels injected into the flare stack or header.

One of the primary advantages of flares, besides having low capital and operating costs, is the ability to handle a wide range of flow rates (Grossel, 1986). The lowest flow is the minimum flow which will prevent air incursion downward into the stack tip. The highest flow corresponds to sonic velocity, assuming sufficient pressure is available. However, the thermal radiation to people at grade or on surrounding elevated platforms and the possibility of flameout increase as the throughput increases. These factors need to be evaluated if the throughput is to be increased beyond the original design limit. Procedures for determining the diameter of a flare stack, tip diameter, and stack height are given in the literature (API, 1982, pp. 33-39, 57-60, 64, and 69; Oenbring and Sifferman, 1980; Kent, 1968; Seebold, 1984; Tan, 1967; Tsai, 1985).

For properly designed flares, the flare ignition system is generally very reliable. Failure modes of flares include flame blowout, pilot failure, and loss of auxiliary fuel. If flame failure does occur, the flare becomes a stack. Consideration should be given to making the flare tall enough so that ground-level concentrations of hazardous materials will be no greater than for an acceptable stack (as discussed in the preceding section). For some high-hazard situations, automatic reignition of pilots and backup fuel supplies may be warranted.

To prevent air from entering the stack, the feed header and/or stack can be purged continuously with an inert gas (nitrogen or combustion gases low in oxygen content) or with a noncondensible fuel gas (such as natural gas). The purge rate depends on the stack diameter and the presence or absence of a molecular seal near the top of the flare stack (API, 1982, sections 3.16 and 5.2.2; Reed, 1968; Bluhm, 1964). For an open stack, a suggested minimum nitrogen flow is 10 cubic feet per hour for stacks 10 inches or less in diameter (Bodurtha, 1980, p. 96), increasing to 350 cubic feet per hour for a stack 24 inches in diameter. The flow rate of flammable gas should be high enough to maintain the concentration of flammable materials above the upper flammable limit.

To protect process vessels in which there may be a flammable atmosphere, installation of a flame arrester or water seal near the base of the flare stack may also be warranted, particularly if any of the connected process equipment could operate at negative pressure (vacuum). An alternative method for protecting against flare-stack explosions is to install a suppression system (Peterson, 1967).

A stack flame arrester should have a temperature sensor and alarm to alert operating personnel that flashback to the arrester has occurred and that the arrester is "holding" a flame (Bodurtha, 1980, p. 95). If "holding" is prolonged, the arrester may overheat, causing the flame to propagate upstream through the arrester. A backup procedure should be developed for coping with such fires (such as diverting the flow to a stack via valving upstream of the flame arrester). Failure to provide a flame arrester may result in an explosion pressure of nearly 100 psig in the header, piping, and vessels (catchtank, blowers, etc.), assuming the system is initially near atmospheric pressure.

Water seals should have low-level alarms and must be protected from freezing with a nonreactive antifreeze solution. The antifreeze concentration should be checked periodically.

A catchtank or knock-out drum with demister pads should be installed upstream of the flare stack if necessary to reduce the discharge of liquid, mist, or droplets from the flare stack (Boix, 1985). Emergency flare systems should be equipped with pilot-flame detectors to automatically reignite pilots in the event of flame failure. Alarms should be installed to indicate pilot and reignition failure, and the reignition system should be tested periodically.

Operation and Safety Considerations. The compositions of the gases in all parts (headers, stacks, etc.) of flare systems should be outside of their flammable ranges. Inasmuch as the discharge from the flare is expected to burn in the atmosphere, it may be necessary to dilute some of the streams so as to keep them oxygen-deficient. If a nonflammable gas (e.g., nitrogen) is used for dilution, the operation of the flare may be affected.

Continuity of purge-gas flow and maintenance of oxygen analyzers must be assured. During maintenance, precautions must be taken to assure that no openings are made in the header system while the plant and flare are operating, since natural drafts may draw air into the system, or the operation of a relief valve, for instance, may cause a release of hazardous vapors through an opening. Similarly, piping, fittings, and vessels must be maintained to prevent corrosion and subsequent air in-leakage.

Knock-out vessels must be kept empty to prevent liquid carryover to the flare. Level sensors and alarms may be necessary to indicate potential hazard. Mist eliminators, if used, should be checked periodically to assure proper performance. Water seals must be kept at the proper level to assure protection against flashback. The physical structure of the stack and supporting structures or guy wires must be sound enough to prevent the stack from collapsing. For further guidance on the use of flares, see DeFaveri et al. (1983), DeFaveri et al. (1985), Brzustowski (1977), and Straitz et al. (1977).

3.3.1.5 CATCHTANKS FOR VAPOR-LIQUID SEPARATION

If the vapor to be discharged under emergency conditions is a two-phase, vapor-liquid mixture, the stream is often routed to a vapor-liquid separator, with a catchtank or "knock-out drum" as the most common design (Grossel, 1986; API, 1982, pp. 51-57, 64; Boeije, 1979). There are several types of catchtanks. Descriptions of catchtanks and design information are given in the references (Grossel, 1986; Speechly et al., 1979; Scheiman, 1965; Fair, 1972; Chambard, 1980; Noronha et al., 1982). Phase separation may be needed to assure proper operation of other treatment or disposal systems.

Catchtank Design. The two main criteria for sizing catchtanks are

- sufficient diameter to effect good vapor-liquid separation, and
- sufficient total volume to hold the estimated amount of liquid carryover (Grossel, 1986).

Supplemental design information is given in Appendix D for the design of catchtanks.

For a foamy discharge, the holding volume should be at least 50 percent greater than the liquid volume in the source vessel. Some stable foams may require special design considerations.

Where two-phase flow from a pressurized container (e.g., a reactor) could occur, it should be assumed that the entire contents of the pressurized container will pass to the catchtank unless more detailed two-phase flow calculations show otherwise.

To reduce the size of the separator, it may be operated at higher pressure during blowdown. This can be accomplished by imposing resistance to flow in the vapor discharge pipe by means of a restricting orifice between flanges at the vapor outlet nozzle. Increasing the pressure reduces the volumetric flow by increasing its density; under such conditions, a heavier walled, smaller vessel may be used. Note, however, that the effect of the increased back pressure on the performance of relief devices must be checked.

Discharge piping from the rupture disk or safety valve to the catchtank should be as short and straight as equipment layout permits. The ideal arrangement would have no elbow in the piping run. This arrangement is seldom possible in practice. The next best arrangement is piping with only one long-radius elbow. Where elbows are required (as in relief headers), they should be adequately braced to withstand thrust and bending moments generated by the flowing vapor-liquid mixture (DIERS, 1988).

Operation and Safety Considerations. The following catchtank design features are recommended as good engineering practice (Grossel, 1986):

- a level alarm with a high setting (set low to detect presence of liquid in the catchtank) and a high-high setting
- a temperature sensor and alarm if liquid in the catchtank can polymerize, decompose, or otherwise self-react; and
- a barometric leg of sufficient height, if the discharge from the catchtank goes to a scrubber and the material in the catchtank could react with the scrubber liquid.

Antifreeze protection should be provided if the liquid discharge into the catchtank contains water or an organic substance with a freezing point near ambient temperature. If internal coils are used, coil drainage must be considered; coils should have a generous corrosion allowance and adequate support to prevent mechanical failure. Heating jackets that use the vessel shell as one wall should be avoided because they limit access to the vessel wall for corrosion inspection (API, 1982, pp. 51-57, 64, and Figure B3).

Inerting or purging of a catchtank may need to be provided where flammable vapors are anticipated. In an arrangement where the catchtank is associated with a stack or flare, the purge systems of both units can be designed as one system.

Disposal of Liquid From Catchtanks. It is normal practice to provide a pump to remove the liquid accumulated in catchtanks. Ideally, the liquid would be returned to the process for purification and recovery. An alternative would be to use any combustible liquid as fuel (after dewatering), with any aqueous phase passing through waste-treatment facilities (to remove toxic or other hazardous materials) prior to disposal.

Where an organic liquid may accumulate in a catchtank and float on an aqueous phase, facilities for continuous removal or intermittent skimming should be considered. Continuous addition of liquid and overflow of the water to a drain can be used, if the lighter phase is immiscible. Provisions for periodically raising the level of the aqueous phase to force the lighter fluid through a skimmer connection may be permissible depending on the toxicity of the aqueous phase. The density effects of two phases on level sensing must be taken into account as well.

3.3.1.6 INCINERATORS

Incineration includes burning in boilers (steam generators), furnaces, kilns, and other fired or catalytic equipment. In continuous operation, recovery of fuel values in vented gases, vapors, or liquids is practical. Incinerators are generally unsuitable for emergency disposal of gases and vapors vented from process equipment since allowable incinerator firing rates are usually not high enough to meet emergency needs. Also, incineration (or oxidation) of noncombustible materials may not significantly reduce the toxicity of some materials. However, direct feeding of hazardous materials into a firebox may be practical if the relieving, venting, or transferring rate does not vary widely. If sudden large flows could occur, holdup tanks or gas holders can be used to provide a controlled flow.

Incinerator Design. If the concentration of flammable vapors is typically well below the lower flammable limit (because of the presence of noncombustible gases and water vapor), incineration may be a practical, albeit slow, method of disposal. The incinerator design includes the required temperature range and residence time to achieve the desired destruction efficiency (Bonner et al., 1981). Good mixing of the normal fuel and combustion air with the material to be incinerated is essential. Typically, natural gas or fuel oil is used in incinerators to maintain temperatures. The material to be incinerated is usually injected into or around the fuel flame.

A flame arrester should be installed near the burner in the waste-disposal piping to prevent flashback in the event that the concentration of flammable vapors in the air exceeds the lower flammable limit. Also, a temperature sensor with an alarm should be installed downstream of the flame arrester to detect "flame-holding" at the arrester (Bodurtha, 1980, p. 95). A backup procedure to deal with such fires should be developed, such as diverting flow to a stack via valving upstream of the flame arrester.

Operation and Safety Considerations. Although incinerators are routinely used to dispose of liquid waste, the heat content of most organic material precludes burning at rates high enough to accommodate emergency disposal. Transfer of liquid or liquefied gas to holding tanks, with subsequent incineration at low firing rates, is a more practical method of mitigating leakage.

3.3.1.7 ABSORBERS

If only a small amount of material is to be recovered and the emergency transfer rate is not high, absorption in a nonreactive medium may be practical. Water is commonly used for inorganic liquids and vapors and for water-soluble organic materials. Oils are used for organic liquids and vapors. A consideration in choosing an absorbing liquid may be the practicality of recovering the desired material by distillation or vaporization, extraction, or crystallization.

Absorption in a combustible heavy oil may be practical for disposal of vapors, with subsequent incineration of the oil. A packed column with countercurrent flows could be used (Perry and Green, 1984, pp. *16*-6, *18*-19, *26*-24). Precautions are needed against flammable mixtures in the collection piping and absorbers because the concentration of a combustible mixture is likely to pass through the flammable range while in the absorber column. Such precautions may include prevention of air in-leakage with inert gas.

Design and operational considerations for absorbers are similar to those for scrubbers. An absorber can be viewed as a specialized type of scrubber.

3.3.1.8 ADSORBERS

Adsorption of vented vapors may be practical in a limited number of emergency situations. This method is of interest when attempting recovery of relatively small quantities or low concentrations of leaking material which would otherwise be lost (Goelzer, 1983; Perry and Green, 1984, p. *26*-25).

The most widely used adsorptive material is activated carbon, which is a good adsorbent for a great variety of organic compounds. However, there have been reports of fires when activated carbon is used for oxygenated compounds such as ketones and aldehydes (Naujokas, 1985). Other adsorbents are silica gel, alumina, and molecular sieves, but water vapor may be adsorbed preferentially onto these adsorbents. Some adsorbents may "catalyze" ignition of flammable vapors, polymerization of monomers, or decomposition of unstable, self-reactive materials. Thus, thermodynamic studies and/or testing may be needed to determine the safety of proposed vapor/adsorbent systems. These systems may need to be tested for selectivity, capacity, and ability to recover the vapor if they are to be effective in an emergency.

The velocity of the gas or vapor stream through the bed of adsorbent should be less than 100 feet per minute, and the temperature of the bed during adsorption should not exceed 50°C (Goelzer, 1983). Precoolers or intercoolers may be needed to prevent excessive temperature in the adsorbent. These requirements, together with the calculated quantity of adsorbent material needed to store the depressured vapors, may make adsorption impractical. Other considerations are pressure drop and regeneration capability.

3.3.1.9 CONDENSERS

One method of collecting vapors being vented from leaking equipment is to cool the vapor below the dew point and recover the condensate. For materials with boiling points greater than 80°F, water is usually used as the cooling medium. For low-boiling-point materials (brines down to 0°F and fluorocarbons down to -100°F), refrigeration systems are used. Too much differential can lead to shock cooling and mist generation (Crocker et al., 1985). Significant amounts of vapor remaining in the effluent can be handled by means of scrubbers or flares.

Contact condensers use simple, flexible, inexpensive equipment to bring the vapor stream into contact with the coolant, with the coolant and vapor temperatures closely approaching each other (Perry and Green, 1984, p. 26-30). These condensers can be used if large amounts of coolant can be made available reliably, if the contaminant to be removed is not soluble in the coolant and phase separation can be used to recover the contaminant, or if the contaminant is soluble in the coolant and the cost of recovery by distillation is not too high.

The design of shell-and-tube heat exchangers (Perry and Green, 1984, pp. 10-24, 26-30; Kern, 1950) is based on temperature difference, available coolant flows, and maximum condensing duty (vapor rate and thermodynamic properties). They are used if the coolant and vapor must be kept apart to avoid reactions, if contamination of the coolant would pose a hazard, or if the cost of recovery by distillation is high.

3.3.2 Emergency Process Abort Systems

Emergency Shutdown

Quench Reaction

Kill Reaction

Dump Unit

Process controls can be used to bring an incipient emergency situation under control before a hazardous event occurs, or to shut down a process safely when a potentially dangerous situation arises. Examples of emergency controls are dump systems, quench systems, and leak-sensing trip systems. These controls are most commonly employed in reactive chemical systems.

A *dump system* is a means of diverting a reactive mixture to a safer location which is generally larger than the normal reaction volume. "Dumping" of the mixture may be tripped by detection of, for example, a high temperature deviation which could lead to a runaway reaction if not controlled. The dumping of the process unit is typically done by remote actuation of quick-opening valves and downward flow of the process mixture into a dump tank. The dump tank could be aboveground, below grade, or a "piggyback" arrangement where the tank holding the reactive contents is immediately on top of a dump tank with the dump valve close coupled in between. Other configurations are also conceivable. Actuation of the dump system tends to suppress any reaction in progress due to the increased effective volume of the system resulting in a lower pressure and temperature. Design considerations include time required to dump the system, reliability of the controls and valves, possible leakage of water or process material into the dump tank, and overpressure protection on the dump tank as well as the process unit.

Quench systems have a similar function to dump systems, as they are generally employed in the processing or storage of reactive chemicals or mixtures. A quench system involves either the injection of a material into the process or diverting the process to a location where a quenching substance is already located. Quench systems can act to slow or stop undesired reactions by one or more mechanisms: (1) inhibiting or "killing" the reaction by either changing the chemical reaction or deactivating a catalyst, (2) cooling the process material to below the dangerous temperature with a colder fluid, or (3) diluting the process material. Note that quenching mechanisms can also be employed in dump systems, for example by maintaining a level of water in a dump tank to cool and dilute a dumped reaction mixture. When designing a quench system, sufficient space should be maintained in the reactor or tank to handle the combined volume. If an inhibitor or diluent is injected into a vessel, the pressure developed by the inhibitor/diluent feed pumps or padding gas must be sufficient to overcome the pressure developed by the reaction. Redundancy of process controls and/or actuating valves may be necessary to achieve sufficient reliability for the quench system when needed. In addition, merely dumping the inhibitor or diluent into the reaction vessel may not guarantee effective action. It may be necessary to provide a

means for mixing the additive into the reactive material. Consideration should be given to services failures which may defeat any normal
mixing. For instance, agitation loss in a polymerization reactor may
not only initiate a runaway reaction, but also preclude sufficient mixing if a quench system is depended upon for emergency process control.

Leak-sensing shutdown systems are designed to be actuated when
loss of process containment is detected. The detection may be by
means of vapor sensors or analyzers, or by measurement of process
variables such as pressure and flow. Vapor detectors are discussed in
Section 5.1. More sophisticated measurements such as rate of pressure drop are sometimes employed; detection of pipeline leaks by
mass balances has proved valuable.

Computerized automation of emergency abort systems is relatively
limited to date, but may become more common once the reliability of
such systems is proven. One common example of computerized systems is combustion safeguard systems. A significant advantage of
computerized systems is that complex logic can be employed where
multiple sensors are used in order to minimize the occurrence of false
trips while at the same time ensuring a high degree of reliability. For
example, a computerized emergency control system might be programmed to initiate a certain emergency action such as shutting off
pumps and closing isolation valves when (1) any two adjacent vapor
detectors in an array around the process register above a given vapor
concentration plus (2) any other leak-sensing device (such as a loss-of-
pressure sensor) is tripped. Computer-aided emergency response
measures could be initiated at the same time.

3.3.3 Emergency Isolation of Leak/Break

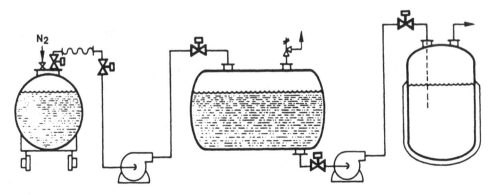

Valves installed between potential leak sites and major sources of inventory are a practical and effective means for limiting the quantity of

material that could be released from piping system leaks. There may be other considerations for installing such valves, such as preventing aggravation of fire or preventing the spread of contaminants in process materials. Other devices (such as positive-displacement pumps, check valves, and excess flow valves) also can serve to limit the quantity of material released. Rupture or leakage at a process vessel will not, of course, benefit from such isolation devices; leaks from vessels can only be isolated using methods such as described in Section 4.7.

3.3.3.1 ISOLATION DEVICES

Automatic Valves Used for Isolation. Spring-loaded quarter-turn valves or gate valves with pneumatic actuators are commonly included in a plant design for isolation, although they may be included for other purposes as well. Block valves installed for isolation should be closely coupled to the process vessels in the event of leakage at the control valve (Lees, 1980, pp. 772 and 773) or connecting piping. Head tanks in particular should be fitted with isolation valves because failure of downstream equipment otherwise could result in release of the head-tank inventory at high pressure (equivalent to the hydrostatic head plus any pad pressure). Long pipelines also should have isolation valves installed to limit the amount of material that might be lost if a break occurs.

Valves used for isolation are often specified to be "fail safe"; that is, they are designed to close on loss of instrument air or electric power at the valve actuator. This is an effective means of increasing the reliability of a remotely operated valve with respect to closing on demand. Backup power or air can be provided if closure on failure of plant supplies cannot be tolerated (as in some cracking furnaces). Air reservoirs should be sized to provide at least two full closings of a valve (Sonti, 1984).

Manual Valves Used for Isolation. Hand-operated emergency shutoff valves can be used provided that they are readily accessible (Lees, 1980, p. 229) to personnel wearing self-contained breathing apparatus (preferably at ground level) and that they are closed and opened frequently to assure proper operation. The risks of sending persons into leak situations to close manual isolation valves need to be considered when including hand-operated valves in a design for leak isolation. The additional risk to personnel together with the additional time required to put on self-contained breathing apparatus often make remote isolation worth the additional expense of hardware installation and testing.

Check Valves. Check valves also can be used as an inexpensive means of preventing backflow from a large container toward a leak, particularly in lines from flexible hoses or articulated arms toward storage tanks. However, check valves are notoriously unreliable in many applications, often because they are not inspected at a suitable frequency. To improve the expected performance, the disk or flapper in such check valves can be assisted by a spring (spring-loaded) or by gravity (lift checks) instead of merely employing a swing device (swing checks). Check valves installed for process-safety purposes can be installed with flanged connections to facilitate removal for periodic inspection.

Excess-Flow Valves. A special type of check valve is the excess-flow valve. The excess-flow valve automatically blocks the flow path if a predetermined flow rate is exceeded, and may be especially effective if used at the outlet of a pressurized vessel. Installing the excess-flow valve inside the vessel itself allows protection against an outlet-nozzle or outlet-flange failure in addition to a downstream transfer line failure; however, inspecting and testing are usually more difficult for such an installation. Liquid eduction pipes in tank cars and tank trucks handling chlorine and other liquefied gases are equipped with rising-ball excess-flow valves. An excess-flow valve cannot be used, however, where insufficient pressure drop is available to actuate the valve, where the process flow rate is close to the maximum rate attainable upon line break, or in services which may interfere with the proper operation of the valve, such as a solids-containing or corrosive stream.

Pumps Used for Isolation. Remote shutdown of pumps, especially if they are positive-displacement pumps, can reduce the rate and duration of leakage. However, pumps should not be solely relied upon as isolation devices, because pumps do not generally provide positive shutoff of flow, and the pump itself may be the source of the leak.

Breakaway Couplings. "Breakaway" couplings can act to isolate parts of transfer systems by shutting off flow if connections at flexible hoses are broken. Such couplings are designed to separate into two halves via frangible bolts if a high stress is applied to the connection (e.g., if a tank car is moved during unloading). Each half of the connection contains a valve which automatically closes to prevent loss of fluid (TCE, 1984). Breakaway couplings are presently in use in many types of transfer and unloading operations involving petroleum products and other hazardous materials.

3.3.3.2 REMOTE ISOLATION

Isolating devices which can be operated from a safe, remote location (e.g., a control building) are the preferred operating control method (Kletz, 1975; Robinson, 1967). The control building should have a reliable fresh-air supply (from two widely separated locations) or should be provided with fresh-air intakes (that can be quickly closed) and airline breathing masks and supplies for the control-room operators (Chemical Industries Association, 1979). To warn the control room operators of hazardous vapor conditions outside, toxic-vapor and/or flammable-vapor detectors should be installed in the air intake ducts (API, 1979).

Switches for actuating isolation valves and shutting off pumps should be neither too close to the point of potential leakage nor too far away. Two push buttons are preferred, with one about 30 feet away and the other in the control room (Kletz, 1975).

Automatic actuation of isolation valves can be accomplished by using polyethylene tubing for air-to-open valves so that any fire in the area would melt the tubing and close the valve. Automatic shutoff of valves upon movement of tank cars can be accomplished by attaching a magnet to the tank car, with the magnet connected to a microswitch by a strong, light cord so that tank-car movement opens the microswitch and closes the valve. Other methods, such as aligning a light beam source on the tank car with a photocell on the unloading platform, have also been employed. Another remote shutoff arrangement which has been employed is a set of remotely operated actuators which quickly attach to a tank car's angle valve handles.

3.3.3.3 INSPECTION AND TESTING OF ISOLATION DEVICES

Block valves used in critical isolation service (and their actuating systems), as well as other types of automatic isolation devices where feasible, should be tested at frequencies sufficient to assure an acceptable probability of proper operation on demand. The desired test frequency for an isolation system may be determined by operating experience and/or quantitative risk assessment. If there is no guidance from operating experience and if a quantitative risk assessment is not made, some authorities suggest a test frequency of once a month (Robinson, 1967; State of New Jersey, 1985); in many situations this could be extended to once a year if there is favorable test experience and low assessed risk.

Testing of the isolation devices needs to be a complete functional check, which includes switch operation through actuation of the final control element (valve, pump, etc.) for remotely tripped systems. Testing under simulated emergency conditions will help demonstrate whether reliability and response times of personnel and equipment are adequate. Check valves and excess-flow valves may not be suitable for some critical isolation services, because it may not be practical to test them without removing them from the piping.

The ability to inspect and test isolation devices needs to be considered at the design stage. "Hold" switches in the field, mechanical stops to limit valve travel (Chemical Industries Association, 1979), or bypass valves (closely attended while open) can be installed to allow the functioning of valves to be checked while the plant is operating. Special procedures generally are required to ascertain proper performance of pump shutoff systems during plant operation. (Some companies now forbid any field bypasses, since these practices can be abused and render critical emergency systems inoperable. Other plants allow field bypasses with special authorization and/or control room display of status.)

3.3.4 Emergency Transfer of Materials

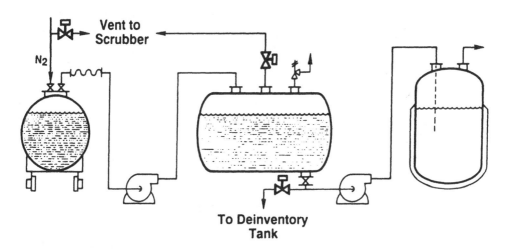

To Deinventory
Tank

Transferring fluids from leaking containers or pipes reduces the duration and possibly the rate of a leak by reducing the driving force of the pressure at the leak. To reduce the leak rate, the pressure at the point of leakage can be reduced (Lees, 1980, pp. 313 and 710), the fluid can be removed from the leaking equipment, or the level of liquid can be lowered below the point of leakage. These objectives can be achieved by releasing the gas or vapor that is providing the pressure to an emergency disposal system or stack or by transferring the liquid to a disposal or recovery system. Any input to the leaking piping or equipment should be shut off to avoid aggravating the leakage. In the event of a process upset or fire, the leak rate should first be reduced, then the inventory should be reduced.

Considerable forethought and planning are necessary if leaking fluid is to be contained and/or recovered rather than destroyed. If the process has been designed to be flexible, that is, if it incorporates oversize piping and pumps, versatile piping interconnections and valving, and "extra" storage tanks, condensers, absorbers, etc., it may be possible to transfer leaking material to storage in an emergency. Even in a flexible process, however, operating personnel will need a certain amount of preparation before they will be able to make the transfer fast enough to reduce the rate or duration of the leakage. If the transfer time is prolonged, other mitigating methods should be used.

When large amounts of material are handled in equipment where leakage is likely, there should be some means of transferring the material to containers that are at lower pressure or elevation if leak isolation is not possible. If these means of transfer are available, all that is required is to determine the proper flow paths to vessels of adequate capacity, determine which pumps or compressors must be put into operation, and periodically test the transfer procedures. Generally, after the transfer, some residual pressure or material will remain and will have to be recycled or disposed of by destruction (flaring, incineration, or reaction), dispersion, or placement into containers for chemical treatment.

More typically, there are no provisions for transferring or receiving materials during leakage. If this is the case, the process or operations may be altered to leave a tank empty so that it can accept the contents of leaking equipment. Additional pumps, piping, and valves can then be installed to stop flows and allow transfer between vessels.

3.3.4.1 TRANSFER OF VAPOR/COVER GAS TO REDUCE DRIVING PRESSURE

Pressure of Noncondensible Gas. The first priority for a leaking vessel should be to depressurize it in order to reduce the release rate. If pressure is provided by a pad of air or other noncondensible gas, the source of pressure should be shut off and the pressure should be vented (preferably through a remote-operated valve) to a safe location. If the pad gas is contaminated with toxic or flammable vapor or mist, it may be necessary to route the gas to a scrubber, flare, incinerator, or tall emergency vent stack.

Vapor Pressure of Liquid at Elevated Temperature. If the vapor pressure is a result of the liquid in the vessel, venting the vapor will not reduce the pressure rapidly because the liquid will boil and generate additional vapor. Eventually the liquid will cool to the normal boiling point; then, the pressure at any leak location below the liquid level will become the liquid head, and the pressure at any leak location above the liquid level will approach zero. If embrittlement from chilling is likely to occur because of emergency venting, low-temperature alloys should be considered in the system design.

Vessel Weakening from Exposure to Fire. If fire exposure is causing release of a nonflammable toxic vapor, or might be expected to cause such a release, it would be desirable to reduce the pressure in the container or process vessel (API, 1976; API, 1982, sections 3.16, 5.2.2; Chiu, 1982; Klassen, 1971; Sonti, 1984; Wells, 1980, p. 190). This is particularly true if fire is impinging on the container above the liquid level, because the heat-dissipating capability of vapor is much less than that of liquid and the metal above the liquid level may become overheated and weakened. In such cases, the container may rupture even though vapor may be venting from the relief device (valve or disk). Then, if the temperature of the liquid contents is above its atmospheric-pressure boiling point, container rupture may be followed by a boiling-liquid-expanding-vapor explosion, or BLEVE (Reid, 1978).

The most effective means of protecting against vessel weakening in a fire situation is to cool the vessel with water. Depressurizing the contents via connecting piping may also be possible. Avoidance of a vessel being engulfed in a pool fire is obviously a preferred protection measure; the pooling of flammable liquids under tanks can be avoided by diverting flammable material away from the process. Dikes and curbs to accomplish the diversion of flammable liquids are discussed in Section 3.4.3.

Emergency Depressuring System. For emergency depressuring, valves can be installed parallel to the relief valve provided for overpressure protection (Klassen, 1971; Sonti, 1984; Wells, 1980, p. 190); thus, the emergency vent valve and the relief valve both vent into the overpressure venting system (flare, scrubber, stack, etc.). However, larger diameter piping (as compared to the overpressure protection pipe diameter) may be needed to accommodate the required vapor flows at lower pressure.

To adequately protect from fire, an emergency vapor depressuring system must have adequate venting capacity. The references give details of depressuring systems for fire exposure (API, 1982, sections 3.16, 5.2.2; NFPA, 1987, paragraph 2-2.5 and Appendix A). Because depressurization must continue for the duration of any fire, the electrical or pneumatic system should be fireproofed so that control valves remain operable during a fire.

3.3.4.2 TRANSFER OF LIQUIDS TO REDUCE INVENTORY AVAILABLE FOR RELEASE

If a vessel is leaking liquid, the depressurization should be accompanied or followed by stopping inflow to the vessel and by transfer of the liquid contents to another container, if possible. For leaks of liquid at temperatures above the boiling point, lowering the liquid level below the point of leakage can significantly reduce the release rate, since a greater mass flow can be released from a flashing liquid stream than from an all-vapor release through the same hole size.

If the container is fitted with a valve at the bottom, it may be practical to drain the contents to another vessel at a lower elevation or pressure or (for liquids with low volatility) into basins for later recovery or treatment. If a pump is installed in the drain line, it may be practical to transfer the contents to another vessel at an equal or higher elevation or pressure. In either case, it will be necessary to ensure that the receiving vessel has sufficient capacity. Depending on the potential hazard and an evaluation of the time in which vessels could be emptied, it may be desirable to keep some storage vessels empty or at low levels for such emergency transfers.

Vacuum trucks can be used for emergency unloading of tanks (or tank cars) containing liquid. These trucks are similar to those used in septic-system cleaning.

General Considerations. It is not possible to pump a liquid at its bubble point from a position above the liquid level, and a syphon will not operate with such a liquid. Thus, a liquid at its bubble point can only be unloaded (1) through a bottom connection, (2) through a top connection after pressurizing the vessel contents, or (3) with a submerged pump.

Valve Requirements. Valves used to drain or vent down equipment are typically single-seated, tight-shutoff valves that are spring loaded and operated pneumatically (by means of a diaphragm or piston) or by an electric motor, quick opening and readily accessible for maintenance and testing (Sonti, 1984). Careful analysis is needed to determine whether the valve should fail open or closed upon loss of instrument air or electric power. Some valves may require backup systems--a "captive" air supply for air pistons, for example. Control valves can also be provided with air cylinders and check valves in the supply lines, with the capacity of the cylinder, for example, sufficient to stroke the valve twice. Another approach to achieving greater reliability is to use dual utility sources, such as instrument air and backup nitrogen with the air and nitrogen lines following different paths to the valve.

3.4 SPILL CONTAINMENT

Engineering design approaches to mitigation can include methods of containing a spill once the primary process containment is breached. Spill containment can serve to prevent escape of the material to the environment or to reduce the vapor release rate by limiting the extent of the spill.

3.4.1 Double Containment

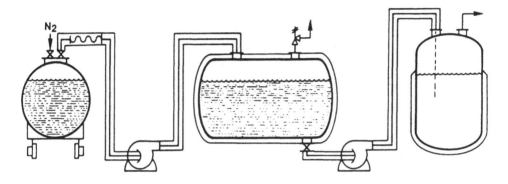

One method to provide spill containment is to essentially double the existing containment provided by piping and/or vessels. Double-walled (concentric) piping and double-walled tanks are discussed in Section 3.1.1 as an engineering design approach to enhance plant integrity.

3.4.2 Enclosures and Walls

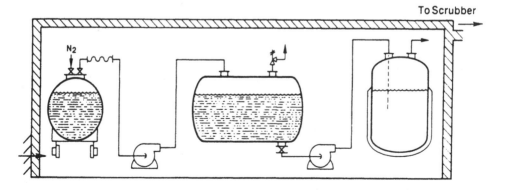

To contain toxic releases in congested areas, secondary containment may be necessary around tank car and tank truck loading and unloading facilities since these locations may account for a large proportion of releases. Secondary containment can be provided by a building or shed. Typically, the building or shed would have exhaust ventilation, actuated by vapor-leakage sensors and directed to a flare or through a scrubber to a tall stack. It is advisable for operating personnel to carry breathing protection (e.g., fresh-air breathing masks with long hoses or self-contained breathing air) at all times while inside an enclosure where hazardous material is present. If the toxic material is also flammable, attention must be paid to the possibility of ignition. For more information on enclosures, see *Guidelines for Storage and Handling of High Toxic Hazard Materials* (AIChE-CCPS, 1988).

Some plants where toxic materials are handled are designed to contain any leakage within the operating buildings. Exhaust ventilation air is scrubbed, flared, or vented through a tall stack. To ensure proper operation of the scrubbing system, redundant controls and leak sensors should be provided.

Walls can aid in diverting heavier-than-air vapors to a safe area (Lees, 1980, pp. 469 and 753) or in temporarily containing vapors in an area where access can be controlled or prevented until countermeasures can be taken (such as spraying the cloud with water) or until downwind areas can be evacuated. This method of containment must be very carefully analyzed before being employed, because of the entrapment dangers it may pose to on-site personnel. Emergency exits would need to be provided through the walls to safe locations.

3.4.3 Dikes, Curbs, Trenches, and Impoundments

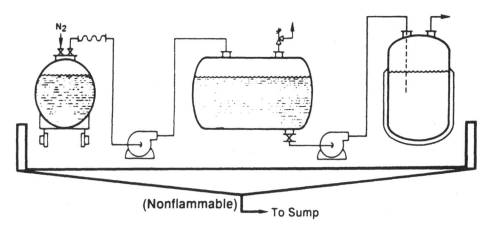

The area covered by a spill can be limited by dikes, curbs, trenches, and impoundments. For spills of toxic liquids, these devices can reduce the rate of evaporation or volatilization. For spills of flammable liquids, these devices reduce the extent of any fire as well as the rate of evaporation or volatilization. For spills of liquefied vapors (i.e., materials which are handled above their atmospheric boiling points), the objectives are similar, but the degree of containment and reduction in vaporization are limited because the liquefied vapor will flash or boil upon contact with the ground or structure.

It is common practice to surround tanks containing flammable liquids with dikes. It is also common to provide slopes, curbs, or trenches leading to impoundments or basins for remote collection of spills (NFPA, 1986; NFPA, 1987, paragraph 2-2.3; Lees, 1980, pp. 220, 239, 525, 753-755, 767, 772, 773, 816). Even if a dike is not drained to an impoundment, it is still good practice to slope the dike away from the tank, to reduce fire exposure to the tank and to minimize the surface area of a spill, which will in turn reduce the rate of vaporization of the spilled material (Harris, 1986). Many diked areas are unnecessarily flat and level, which allows a spill of volatile liquid to spread over the entire surface of the level area.

A dike or impoundment facilitates covering the spill with foam to retard evaporation and reduce the probability of ignition. However, if the flammable liquid ignites, a dike tends to surround the tank with fire. Thus, a diversion system can be provided to take the material away from the tank and thus reduce the intensity of fire at the tank. Fire in the impoundment can be attacked with foam or dry chemical. Also, impoundments can be designed to keep the surface area low which will reduce the intensity of the fire and facilitate fire fighting.

Containers of toxic materials can also be surrounded by dikes to reduce the rate of vaporization and thereby limit the distance dangerous concentrations of the toxic material can attain. Providing curbs approximately 18 inches high within dikes containing several tanks (Lees, 1980, pp. 753, 754) can further reduce the area of vaporization and aid in covering a spill with foam or other medium to inhibit vaporization.

Full-height dikes have been provided for cryogenic ammonia and liquefied natural gas (generally in Europe), with 28-inch-high dikes for ambient-temperature chlorine (COVO, 1982). Full-tank-capacity dikes have been provided for materials with boiling points above or modestly below ambient temperature, such as hydrogen fluoride, methyl mercaptan, hydrogen cyanide, and phosgene.

As a general rule, the height of a dike has been established by the volume of the material in the tank, calculated such as in equation (3-1):

$$h = \frac{(\pi/4)D^2H}{A - (\pi/4)D^2} \tag{3-1}$$

where h is the height of the dike, D is the diameter of the vertical cylindrical tank in the dike, H is the height of the tank, and A is the area bounded by the dike, all in consistent units. This does not take into consideration that if the tank were to burst, the momentum of the contents might cause overflow of the dike wall. As a result, a dike might have to be somewhat higher to ensure containment in the event of a severe tank rupture. Local codes or governmental regulations may have specifications on dike heights or volumes for certain materials.

An additional consideration when designing a dike for spill containment is the material of construction. Not only should the dike material be compatible with the process material (as discussed in Section 6.3) and be constructed to minimize surface area, but thermal factors can also be considered. Some companies have used low-density concrete for dike construction, particularly where the ambient temperature is above the boiling point of the process material (i.e., where pool boiling is likely if a spill occurs). The low-density concrete reduces the rate of heat transfer to the spilled material and thus reduces the vaporization rate. It may also be possible to use an insulated liner on a dike or collection pit.

REFERENCES

AIChE-CCPS, 1987: *Guidelines for Use of Vapor Cloud Dispersion Models*, prepared by S. R. Hanna and P. J. Drivas for American Institute of Chemical Engineers-Center for Chemical Process Safety, New York.

AIChE-CCPS, 1988: *Guidelines for Storage and Handling of High Toxic Hazard Materials*, prepared by Arthur D. Little, Inc. and R. LeVine for American Institute of Chemical Engineers-Center for Chemical Process Safety, New York.

API, 1976: Recommended practice for the design and installation of pressure-relieving systems in refineries, Part I--Design. *API Recommended Practice 520*, 4th Ed., American Petroleum Institute, New York, pp. 17, 18 and Sections 6.2, 8.

API, 1979: *Safety Digest of Lessons Learned*, Publication 758, Section 2, American Petroleum Institute, New York, 157.

API, 1982: Guide for pressure-relieving and depressuring systems. *API Recommended Practice 521*, 2nd Ed., American Petroleum Institute, New York.

ASME, 1974: *ASME Boiler and Pressure Vessel Code*, Section VIII, American Society of Mechanical Engineers, New York.

Baumeister, T. (ed.), 1979: *Marks' Standard Handbook for Mechanical Engineers*, 8th Ed., McGraw-Hill, New York, 6-106.

Bluhm, W. C., 1964: Safe operation of refinery flare systems. *Fire Protection Manual for Hydrocarbon Processing Plants* (C. H. Vervalin, ed.), Gulf Publishing Co., Houston.

Bodurtha, F. T., 1980: *Industrial Explosion Prevention and Protection*, McGraw-Hill, New York.

Boeije, C. G., 1979: Flare relief systems. Inst. Chem. Eng., North West. Branch Meet., 6.17.

Boix, J. A., 1985: Emergency flare systems--Some practical design features. *Plant/Oper. Prog. 4*, (1), 222.

Bonner, T. A., et al., 1981: *Hazardous Waste Incineration Engineering*, Noyes Data Corp., Park Ridge, NJ.

Brzustowski, T. A., 1977: Flaring: The state of the art. *Loss Prev. 11*, 15.

Chambard, J. L., 1980: Designing processing pressure relieving systems for safe performance. *Int. Symp. Loss Prev. 3rd* (Basel, Switzerland), 1230.

Chemical Industries Association, 1979: Process plant hazard and control building design. *Guidelines for the Design of Control Buildings Subject to Toxic Gas Hazard*, Appendix III, Chemical Industries Association, London.

Chiu, C. H., 1982: Apply depressuring analysis to cryogenic plant safety. *Hydrocarbon Process., 61* (11), 255.

COVO, 1982: *Risk analysis of six potentially hazardous industrial objects in the Rijnmond area, a pilot study*, Dutch Commission for the Safety of the Population at Large (COVO), D. Reidel Publ. Co., Dordrecht, Holland, 130, 151, 153, 184, 288.

Crocker, B. B., et al., 1985: Air pollution control methods. *Kirk-Othmer Concise Encyclopedia of Chemical Technology*, John Wiley & Sons, New York, 46.

DeFaveri, D. M., et al., 1983: *Int. Symp. Loss Prev. 4th* (Harrogate, England), *Inst. Chem. Eng. Symp. Ser. 82*, G-23.

DeFaveri, D. M., et al., 1985: Estimate flare radiation intensity. *Hydrocarbon Process, 64* (15), 89.

DIERS, 1988: *DIERS Project Manual*, Design Institute for Emergency Relief Systems, American Institute of Chemical Engineers, New York (draft manual available in 1987; publication in 1988).

Fair, J. R., 1972: Designing direct-contact coolers/condensers. *Chem. Eng. 79* (12), 91.

Feldman, R. J. and S. S. Grossel, 1968: Safe design of an ethylene plant. *Loss Prev. 2*, 131.

Fenlon, W. J., 1984: A comparison of ARC and other thermal stability test methods. *Plant/Oper. Prog. 3* (4), 197.

Fisher, H. G., 1985: DIERS research program on emergency relief systems. *Chem. Eng. Prog. 81* (8), 33.

Gackenbach, R. E., 1960: *Materials Selection for Process Plants*, Reinhold, New York.

Goelzer, B., 1983: Air pollution control equipment. *Encyclopedia of Occupational Safety and Health, 2*, 3rd Ed., International Labor Office, Geneva, 2377, 2384-2388.

Grossel, S. S., 1986: Design and sizing of knock-out drums/catchtanks for reactor emergency relief. *Plant/Oper. Prog. 5* (3), 129.

Hamner, N. E., 1974: *Corrosion Data Survey, Metals Section*, 5th Ed., National Association of Corrosion Engineers, Houston.

Hamner, N. E., 1975: *Corrosion Data Survey, Nonmetals Section*, 5th Ed., National Association of Corrosion Engineers, Houston.

Harris, C., 1986: Mitigation of accidental toxic gas releases. Paper presented at Toxic Release Assessment and Control Seminar, Chemical Manufacturers Assoc., Washington, D.C.

Hoot, T. G., and R. N. Meroney, 1974: The behavior of negatively buoyant stack gases. Paper presented at 67th Annu. Meet. Air Pollut. Control Assoc., Denver.

Hub, L., and J. D. Jones, 1986: Early on-line detection of exothermic reactions. *Plant/Oper. Prog.*, 221.

Jenkins, J. H., et al., 1977: Design for better safety relief. *Hydrocarbon Process., 56* (8), 53.

Kent, G. R., 1968: Find radiation effect of flares. *Hydrocarbon Process., 47* (6), 119.

Kern, D. Q., 1950: *Process Heat Transfer*, McGraw-Hill, New York.

Klassen, P. L., 1971: Loss prevention aspects in process plant design. *Inst. Chem. Eng. Symp. Ser. 34*, London, 120.

Kletz, T. A., 1975: Emergency isolation valves for chemical plants. *Chem. Eng. Prog. 71* (9), 73; *Loss Prev. 9*, 134.

Kohan, D., 1974: The design of interlocks and alarms. *Chem. Eng. 91* (4), 73.

Lees, F. P., 1980: *Loss Prevention in the Process Industries*, Butterworth, London.

Naujokas, A. A., 1985: Spontaneous combustion of carbon beds. *Plant/Oper. Prog. 4* (2), 120.

NFPA, 1986: *Fire Protection Handbook*, 16th Ed., National Fire Protection Association, Boston, *11*-34.

NFPA, 1987: *Flammable and Combustible Liquids Code*, NFPA 30, National Fire Protection Association, Boston.

Noronha, J. A., et al., 1982: Deflagration pressure containment. *Plant/Oper. Prog. 1* (1), 1.

Oenbring, P. R. and T. R. Sifferman, 1980: Flare design: Are current methods too conservative? *Hydrocarbon Process. 59* (5), 124.

Perry, R. H. and D. W. Green, 1984: *Perry's Chemical Engineers' Handbook*, 6th Ed., McGraw-Hill, New York.

Peterson, P., 1967: Explosions in flare stacks. *Loss Prev. 1*, 85; *Chem. Eng. Prog. 63* (8), 67.

Prugh, R. W., 1974: Preferred pilot plant control practices. *Chem. Eng. Prog. 70* (11), 61.

Reed, R. D., 1968: Design and operation of flare systems. Loss Prev. 2, 114; *Chem. Eng. Prog. 64* (8), 53.

Reid, R. C., 1978: Superheated liquid: A laboratory curiosity and, possibly, an industrial curse. *J. Chem. Educ. 12* (2), 60; *12* (3), 108; *12* (4), 194.

Robinson, H. S., 1967: Loss risks in large integrated chemical plants. *Loss Prev. 1*, 20.

Scheiman, A. D., 1965: How to size shower deck baffled towers quicker. Part 1--Tower diameter. *Petro/Chem. Eng. 37* (3), 28. Part 2--Tower tangent length. *Petro/Chem. Eng. 37*, (4), 75.

Seebold, J. G., 1984: Practical flare design. *Chem. Eng. 91* (95), 69.

Slade, D. H., 1968: *Meteorology and Atomic Energy*, U.S. Atomic Energy Commission, Report TID-24190, 404.

Sonti, R. S., 1984: Practical design and operation of vapor-depressuring systems. *Chem. Eng. 91* (2), 66.

Speechly, D., et al., 1979: Principles of total containment system design. Inst. Chem. Eng. North West. Branch Meet., 7.1.

State of New Jersey, 1985: *Toxic Catastrophe Prevention Act*, Assembly Bill 4145.

Straitz, J. F., et al., 1977: Flare testing and safety. *Loss Prev. 11*, 23.

Tan, S. H., 1967: Flare system design simplified. *Hydrocarbon Process., 46* (1), 172.

TCE, 1984: Anon., *The Chem. Eng.* (England), (10), 26.

TNO, 1980: Methods for the Calculation of the Physical Effects of the Escape of Dangerous Material (Liquids and Gases), Netherlands Organization for Applied Research (TNO), Voorburg, The Netherlands, chapter 7.

Trbojevic, V. M. and Y. N. T. Maini, 1985: An approach to the assessment of double containment of refrigerated liquified gas storage tanks under abnormal liquid loads. *The Assessment and Control of Major Hazards, Inst. Chem. Eng. Symp. Ser. 93*, London, 124.

Tsai, T. C., 1985: Flare system design by microcomputer. *Chem. Eng. 92* (17), 55.

Wells, G. L., 1980: *Safety in Plant Design*, John Wiley & Sons, New York.

Wilke, C. R. and U. Von Stockar, 1985: Absorption: *Kirk-Othmer Concise Encyclopedia of Chemical Technology*, John Wiley & Sons, New York, p. 4.

4

PROCESS SAFETY MANAGEMENT APPROACHES TO MITIGATION

In addition to making a plant or process inherently safer and incorporating engineering approaches to mitigation in the plant design, another basic method to mitigate vapor release hazards is by process operating approaches. This includes all of the process risk management efforts taking place at an operating facility. Without the proper risk management elements and attitudes in place, nearly all engineered approaches to mitigation become unreliable. For example, without adequate testing, emergency isolation hardware is likely to be unavailable when needed.

The most important process operating approaches are described in this chapter. An AIChE-CCPS Technical Management Committee document entitled *Chemical Process Safety Management--A Technical Approach* (in preparation) will give further details on the subject of risk management guidelines.

4.1 OPERATING POLICIES AND PROCEDURES

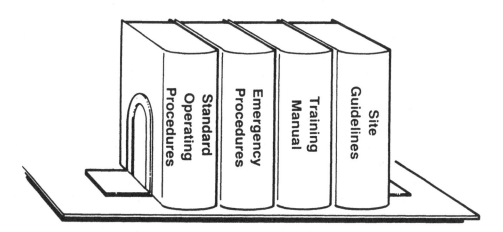

Most major incidents can be traced in part to human error. This error may range from a management system failure to a process operator's failure to follow instructions. One of the most vital management systems required to prevent and mitigate vapor releases of chemicals is a highly visible and well understood Operating Policies and Procedures program.

Policies and procedures are a systematic approach for management to clearly express its expectations for safe operation. Operating policies should clearly state the intentions of management to equate safe operation to other activities ongoing at the site (production, quality control, etc.).

Operating procedures should be action oriented and should be focused on "how to" steps necessary for safe operation. These procedures are usually aimed at outlining the accepted means of safely conducting the operation. Good procedures are simple and clear. They are used as a basis for training. Well-written procedures are clear as to what parts are mandatory and to what parts are guidelines.

Examples of typical procedures include the following:

- *Standard Operating Procedures.* Minimum contents of a standard operating procedure should include a simplified process flow description and diagrams, a control system description, normal operating parameters, shutdown and interlock system functions and rationale, reference to "never-exceed" limits, and the steps and sequences required to safely operate the process. The standard operating procedure is usually a basis of operator training and must be aimed at being usable and understandable by a qualified person.
- *Variance and Exception.* Depending on management philosophy, if acceptable alternatives can be presented for an otherwise mandatory procedure or standard, allowance can be made through a documented method.
- *Communications.* Minimum standards for communication of process status, events in progress, and changes in operating conditions should be established between process operators (especially if shift operations are involved) and the supervision directing the operation.
- *Abnormal Operation or Emergency Procedures.* Scenarios for foreseeable emergencies should be preplanned, and accountabilities for mitigation assigned and rehearsed in advance. These should include foreseeable process abnormalities such as loss of specific utility services, leaks and ruptures, severe weather, and emergency shutdowns. Chains of command and communication must be clear, both inside the chemical plant

organization and with the community. Should evacuation or special sheltering of impacted personnel be required by these plans, the decision-making process must be clearly documented.

- *Start-up, Shutdown, Standby.* Because many incidents occur during transient operations, it is very critical that these unusual operating periods have clear and specific instructions. These instructions can be presented to the process operator in the form of checklists, step-by- step instructions and restrictions, or simple logic diagrams for timely decision making.

- *Critical Equipment Identification.* Equipment critical for safe operation must be defined and identified. Procedures which control the reliability and operability of critical equipment must be provided. These would include: critical safety devices, critical control loops, critical safety trips and interlocks, pressure relief devices, and devices which sense the approach to a "never-exceed" limit and take action to prevent the occurrence of a major hazard. Procedures that allow the bypassing of interlocks must be well documented and must have strict administrative controls.

- *Hazardous Work Procedures.* These identify the steps necessary to safely perform work on process equipment. Typically, these procedures include the steps necessary to bring equipment to a safe state (emptying, flushing, blocking in, disconnecting from associated equipment, isolation, and special tagging) and work permits so that repairs and inspections can be conducted safely.

- *Modifications.* Accountabilities for safety reviews of process modifications must be detailed, as well as authority to begin operations after the modifications are completed.

- *Incident Investigation and Communication.* Incidents should be thoroughly investigated for their basic causes, with actions taken to prevent recurrence. Lessons learned should be shared throughout the organization to improve awareness and knowledge. A trend of increasing frequency of incidents should be taken as a strong negative indicator.

- *Hazard Analysis.* The specific hazards involved in a chemical process must be analyzed for their severity and impact on personnel, the process, and the community. Strategies for controlling, mitigating, and (where feasible) eliminating these hazards must be documented and practiced in the plant operation.

All procedures must be reviewed at specific intervals for accuracy and content. Modifications and changes in the process must also be incorporated into the body of operating procedures. Temporary procedures, variances, and exceptions should all be tested for a specified period for validity and incorporated into the standard procedures once validated.

4.2 TRAINING FOR VAPOR RELEASE PREVENTION AND CONTROL

Training of personnel involved in either the prevention or the control of vapor releases is necessary for an effective program. Knowledge and skills for avoiding incidents is a large part of mitigation. Motivating plant personnel to use this knowledge is an important part of supervision.

Those involved in preventing vapor releases need good operating techniques, documented in operating procedures which have specific "how to" information. These documented procedures form a basis of the training curriculum and should identify the chemical and process hazards and the detailed procedures to control these hazards.

Critical areas for training in prevention of releases include

- process variables and design conditions;
- hazards of process chemicals and intermediates involved;
- process start-up, shutdown, and abnormal or emergency procedures;
- preparation of equipment for repairs and inspection;
- safe handling, storage, and transfer of chemical products;
- control of process modifications;
- equipment critical for safety; and
- critical operating parameters.

Testing of the knowledge received in training is the only way to assure qualifications. Testing should be based on management's expectations for the operating position and must cover the minimum skills and abilities necessary to safely operate the process.

Effectiveness of employees in preventing vapor releases should be a part of annual performance reviews. Periodic, routine safety meetings are an excellent means of maintaining awareness of process hazards and vapor release control.

Additional training and expertise must be provided for those involved in control and mitigation of vapor releases. Scenarios of foreseeable events resulting from loss of hazard control should be developed, and these scenarios can be beneficially used in training operators and emergency response personnel. These scenarios are an excellent means of sharpening the skills and knowledge of emergency response personnel both inside and outside the plant facility. Successful scenario drills include rapid

- detection of the leak or emission,
- alarming of the work force and emergency services, and
- combating the leak itself.

Practical scenarios should be rehearsed routinely. Additional skills are gained by adding complexity. Exercises can range from tabletop role playing to complete on-scene live rehearsals. Knowledge of alarm signal recognition, use of fire-fighting apparatus, spill and leak containment equipment, and emergency notification procedures should all be part of a drill. In each instance, a critique of the exercise should be performed to capture both the good and bad points. These critiques can lead to action plans to improve plant design, emergency equipment, communications, etc.

Emergency scenario training is an excellent way for team building in organizations which may have multiple groups and squads involved. Constructive competition for the best solution to difficult emergency situations has many times led to safer plant operations.

Identifying scenarios which have no obvious, readily implemented solution acceptable to management is an important outcome of hazard analysis. Other solutions which may involve eliminating or reducing the hazard may require systematic risk assessment techniques to assess postulated solutions.

4.3 AUDITS AND INSPECTIONS

Performance standards for the facility operations should be established and used so that inspections are carried out at the specified time and an acceptable degree of compliance is realized. Whether or not these performance standards are achieved should be part of the annual appraisal system.

External audits and inspections can be in the form of a deficiency analysis. The facility is surveyed to determine the plant status on a wide scope of concerns. Examples would be

- adequacy of fire protection equipment;
- records and trends in equipment testing and maintenance;
- adherence to management procedures, policies, and standards;
- training and drills on emergency scenarios;
- compliance with design and fabrication standards;

- overview of the on-site inspection history and follow through;
- materials-of-construction audits;
- examination and reevaluation of the justification for variances and exceptions;
- examination of facility modification documentation; and
- review of past incidents and actions defined to prevent recurrence.

These audits can occur any time during the life of the facility.

The auditing of new processes both during design and final construction for compliance with all standards is a strong contributor to controlling and mitigating vapor releases. Periodic follow-up audits are necessary to see that the facility is properly maintained and operated, and that the equipment and procedures are modified according to accepted standards.

4.4 EQUIPMENT TESTING

Systematic and routine testing for integrity of process equipment and piping is necessary to determine if imperfections are present which threaten loss of containment. These imperfections can be discovered by thorough inspection of equipment in all phases of its life. Testing programs for maintaining the reliability of mitigation equipment are also necessary to ensure the continuing effectiveness of mitigation systems.

Imperfections can be discovered prior to commissioning and start-up when they are due to inadequate fabrication techniques (poor workmanship, poor quality control, wrong materials, etc.), inadequate design for the proposed duty (inadequate pressure or temperature ratings, etc.), physical damage during transit or storage, or defects during assembly (poor alignment, wrong gaskets, insufficient welds, etc.). A variety of testing techniques can be employed prior to start-up, such as dye checking and X-ray testing.

Imperfections also arise from equipment deterioration in service, thus requiring a routine testing and inspection program to detect these imperfections. Deterioration causes include: wear on rotating equipment seals, valve packing, and flange gaskets; internal and external corrosion and cracking; special chemical attack such as crevice crack corrosion or hydrogen embrittlement; erosion or thinning due to flow; metal fatigue or vibration effects; and the effects of transient or long-term operation beyond design limits. These problems may be particularly severe in high stress areas, such as bellows expansion joints.

Imperfections may also arise due to improper maintenance and repair techniques. In addition, modifications may take place which compromise the containment integrity. A maintenance quality assurance program should recognize these potential hazards. Maintenance procedures and repair plans which include prescribed tolerances, precise parts descriptions, exact materials of construction requirements, and assembly instructions reduce the chances of individual error.

A full functional test of control and containment components gives clear and documentable evidence of their ability to contain the process. This test should specify that the input stimulus be as close to the real process input as practical. Examples are

- operation of shutdown and trip devices based on real inputs to the detection system through its logic devices, and verification of the final control element action or movement;
- calibration and spanning of process analyzers using real process fluids of known composition as inputs and verification of the indicated analysis; and
- hydrostatic testing of process containment equipment at the maximum allowable working pressure with zero leakage.

Exposure to abnormal conditions, such as fire or serious leakage of corrosive liquid or vapor, should be followed by close inspection of all process equipment and structures.

Inspection and test procedures should apply to all equipment where failure could lead to injury or property loss. They apply especially to standby equipment such as scrubbers, stacks, and flares which could be needed in the event of plant upset for disposal of hazardous materials. Pressure relief valves can be removed for bench testing and verification of their activation setting. Rupture disks should be inspected periodically for defects, damage, and plugging. In some applications it may be necessary to replace them automatically on a scheduled basis to avoid premature failure from low-level fatigue, stress, vibration, etc.

To be able to make serviceability predictions with confidence, inspections and tests must be intensive and conducted at an appropriate frequency. A philosophy that may be developed is that the inspection or test frequency is increased if any statistically significant or interval-dependent failures are found, and that the frequency may be decreased if several consecutive tests or inspections are passed successfully.

A systematic and routine program of equipment testing can give plant management the data required to predict equipment failure rates. The tracking of the data allows the materials engineer to schedule and perform proper repairs or to recommend changes in operating conditions to minimize further deterioration. In some cases, alternative materials of construction and fabrication techniques may be recommended.

It may be difficult to schedule inspections of the process equipment when they are due. Inspections often involve downtime and the expense of cleaning equipment for entry and examination. Scheduling conflicts and business needs may interfere. Nonetheless, management must assess the need to ensure proper containment of hazardous materials and supply appropriate resources to carry out a sufficiently intensive testing program.

An inspection program should extend to storage tanks, stacks, columns, and other equipment involved in hazardous material service to assure that subsidence or tilting is not occurring. This particularly applies to equipment on filled ground or where a high water table requires extensive use of piles for equipment support.

4.5 MAINTENANCE PROGRAMS

An effective preventive maintenance program for chemical process equipment is required to ensure containment integrity. This program should include technical input, tracking of equipment failure data, and a reporting system that highlights to management deviations from planned assurance programs. Gauges and measuring devices that are used to calibrate pressure relief devices, rotating equipment clearances and tolerances, metal thicknesses, etc. should conform to recognized standards and should be recalibrated as appropriate.

Process equipment and instrumentation that are critical to process safety should be identified and their design documented. Equipment critical for safe operation can be defined as the equipment or systems installed to avoid or mitigate major hazard incidents, or equipment or systems which upon failure could cause a major hazard incident. Examples are

- equipment designed to detect an operating parameter's approach to a "never-exceed" limit and/or designed to take action upon reaching this limit,
- emergency relief systems, and
- equipment especially required to be in service by process safety standards mandated by management directives or regulations.

A comprehensive hazard evaluation can facilitate identification of equipment critical for safe operation.

Maintenance test and repair histories should be kept for process equipment. Frequencies for testing equipment should be specified and a monitoring program should be established to ensure the schedules are kept. By tracking repair and testing histories, imperfections due to either mechanical or chemical action can be trended. Proactive repair and/or replacement of equipment showing an unsatisfactory maintenance history can prevent or minimize material releases to the environment.

Early detection of cracking, corrosion, and metal damage can give the facility management a chance to prevent an incident. Documentation of this deterioration should employ a formalized system.

Adherence to pressure vessel and piping codes and repair standards is another means of achieving maintenance quality assurance. Where repairs are to be performed on code-certified vessels, the regulations dictate that the craftsman and inspector performing these repairs should be qualified for the task. After any such repairs or modifications are made, thorough testing (per the code) of the repair/modification should be performed before the equipment is returned to process service. Procedures that name specific, technically qualified persons as the only ones authorized to specify the type of alterations and repairs made to coded vessels limit the opportunity for unsanctioned modifications.

Maintenance procedures should include detailed disassembly, repair, and reassembly instructions especially for equipment critical for safe operation. Parts descriptions, materials of construction, and alternative parts and materials which either can or cannot be substituted take much of the human error element out of repair activities.

Safe work practices, such as equipment isolation, tag out, lock out, cleaning and flushing, etc., contribute to the prevention of process releases by establishing an administrative layer of defense for loss of containment. This is accomplished by controlling non-operating-personnel access to the process equipment. The probability that the wrong equipment could be accidentally opened, resulting in release of hazardous process materials, is substantially reduced through the use of these procedures.

4.6 MODIFICATIONS AND CHANGES

A change to the chemical process or facility, however small, can inadvertently introduce a new hazard or maybe remove a layer of defense from a known hazard. These alterations may be of sufficient magnitude to result in loss of containment of a hazardous material.

To ensure that the potential for vapor releases does not increase, a good system of checks and balances should be incorporated within the operations area. This would include a safety review of any intended change before finalizing its design, followed by a safety review of the completed change before it is started up. The first review is called a "facility change review," and the second is called a "pre-start-up safety review." Both reviews should include participation by competent persons not involved in designing or implementing the change, so that an independent assessment of its appropriateness and implications is made.

Many small changes (piping, jumpers, addition of drain connections, bypasses, pump seal modifications, etc.) may not be specified by a formal design procedure. Nonetheless, these modifications are still changes and must be thoroughly reviewed for their process safety implications.

A change to a specific process or facility can be defined as any deviation from, or modification to, accepted practices or technology of known risk potential, or operations outside the known "envelope of technology" defined for the processes. Examples of changes include

- introducing a new chemical to the facility,
- using a new catalyst in the process,
- modifying a vessel or piping,
- changing an operating parameter beyond accepted norms,
- restarting an idled facility, and
- making control strategy changes.

The basis of the facility change review is that one or more qualified persons systematically review the changed design for introduction of new hazards. If new hazards are discovered, then the design must be modified to eliminate, control, or mitigate these hazards to levels of risk acceptable to line management.

A facility change review should document that the management in charge of the facility is fully aware of (1) the intention and scope of the proposed change, (2) the documented independent safety assessment of the change, and (3) authorization by the appropriate responsible person. Many companies have a tiered system of facility change approval based on the magnitude of the change--the more significant a change is to the process technology, the more management is involved in approving the change.

Following approval of the facility change review, installation of the modification takes place. If a major construction effort is involved in the modification, there may be a long interval between the facility change review and eventual start-up. During this period, management must be vigilant because the original design is vulnerable to further field changes and well- intended alterations.

Following the end of installation and just prior to commissioning, the pre-start-up safety review should be conducted. Its purpose is to look at the finished facility and again look for new hazards or changes in control of existing hazards. Questions should be asked by the reviewer to determine if the concerns raised by the facility change review were indeed carried out. In addition, the pre-start-up safety review should ensure that any deviations from the intended design are thoroughly analyzed, including fabrication or installation errors and their resolution.

The person(s) conducting the pre-start-up safety review should prepare a written report to management listing their findings and highlighting any deficiencies discovered which must be remedied prior to start-up. Management approval of the review and its finding should be documented.

One element that should always be completed prior to start-up is the updating of standard operating procedures that apply to the modification. The process operators' understanding of the impact of the change can only be achieved through training, which in turn relies on good operating procedures.

4.7 METHODS FOR STOPPING A LEAK

4.7.1 Patching

A direct method for countering leaks is blocking them with plugs, caps, air bags (Hamlin, 1985, p. 7-1), or other materials. The Chlorine Institute has developed three kits (A, B, and C) for temporary repairs to leaking chlorine cylinders, ton containers, and tank trucks, respectively (Chlorine Institute, 1986, p. 20). Similar emergency kits have been developed by manufacturers and users of other hazardous materials. The kits contain patching devices, such as tapered plugs, permanent-magnet patches, strongbacks or clamps attached to permanent magnets, or electromagnets (powered by mobile welding machines) with epoxy adhesives and assorted gaskets. These kits may also contain protective suits and masks.

Before such kits are made a part of a spill emergency response program, it is essential that they be tested for suitability on the containers from which leaks might occur to ensure that

- patches are of suitable size and contour and have adequate seating and gasket-compression force;
- suitable access platforms, ladders, and "cherry pickers" are available;
- equipment can be used by personnel when wearing protective clothing and emergency breathing apparatus;
- magnetic devices are applicable to containers (e.g., stainless steel); and
- electrical codes permit the use of any electromagnets required.

Personnel most likely to use the leak-patching equipment must be trained in its use and made aware of its limitations. They should also be trained to analyze the situation before applying a patch. For example, a "leaking" tank car relief valve may indicate an overfilled tank car or an internal reaction.

4.7.2 Freezing

If a leaking material has a relatively high freezing point and the leak is at a valve or fitting in accessible piping, freezing the leaking material by external cooling may quickly stop a small leak. Embrittlement of construction materials and creation of stresses in the piping from thermal expansion/contraction during freezing should be carefully considered before this alternative is selected.

Carbon dioxide fire extinguishers have been used successfully for materials with freezing points somewhat above the sublimation temperature of carbon dioxide "frost" (-78°C). Presumably, other liquefied frozen gases such as nitrogen (Schelling, 1986) could be used to create plugs of frozen material in piping, thus reducing leak rates.

Care should be taken to avoid asphyxiation when using large quantities of carbon dioxide, nitrogen, etc., especially inside a building or confined area.

Table 4-1 presents freezing point data for liquefied vapors, for comparison with the normal boiling points of carbon dioxide (-78°C) and nitrogen (-196°C).

TABLE 4-1. FREEZING POINT DATA

Pure component	Freezing point (°C)[a]
Acrolein	-87.0
Ammonia	-77.7
Arsine	-116.3
Bromine	-7.2
Carbon disulfide	-111.5
Chlorine	-101.0
Ethylene	-181
Formaldehyde	-92.0
Hexane	-95.6
Hydrogen chloride	-114.8
Hydrogen cyanide	-13.2
Hydrogen fluoride	-83.1
Hydrogen sulfide	-85.5
Methyl mercaptan	-123.1
Dimethyl sulfate	-31.8
Nickel carbonyl	-25
Phosgene	-118
Phosphine	-133
Propane	-187.1
Sulfur dioxide	-75.5
Sulfur trioxide	16.8
Titanium tetrachloride	-25
Vinyl acetate	-100.2

[a] Melting point used where freezing point data not available.

4.8 SECURITY

Industrial security is outside the scope of these guidelines, except to note that inadequate security could impact plant integrity. In several ways, protection of process equipment also protects against actions of intruders. Typical procedures include locating vulnerable processes near the center of the plant. This also provides a degree of isolation for toxic-hazard processes (Lees, 1980, pp. 221, 720), though without precautions it may increase on-site risk. Installing high-intensity lighting and television monitoring for leaks and spills (which deters and could detect intruders), and safeguarding computer control terminals where operating ranges, software overrides, and interlock set points can be changed, may also be appropriate.

REFERENCES

Chlorine Institute, 1986: *The Chlorine Manual*, 5th Ed., The Chlorine Institute, Washington, D.C.
Hamlin, F., 1985: An analysis of the Mississauga chlorine tank car accident. *Proc. 28th Chlorine Plant Oper. Semin.*, 7-1.
Lees, F. P., 1980: *Loss Prevention in the Process Industries*, Butterworth, London.
Schelling, C., 1986: Reducing repair downtime with pipefreezing. *Plant Eng. 40* (15), 55.

5

MITIGATION THROUGH EARLY VAPOR DETECTION AND WARNING

In the event of a release of hazardous vapors to the environment, prompt detection and warning plays a vital role in mitigating the consequences of the vapor release. Detection and warning are necessary to minimize the time required to motivate the best response to the vapor release.

5.1 DETECTORS AND SENSORS

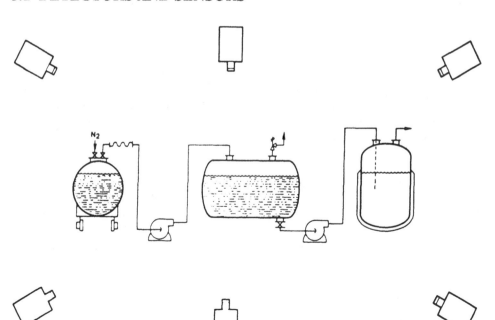

For isolated facilities, vapor detectors should be installed at "critical" locations (Kletz, 1975) to provide early warning of impending hazard. In all plants, vapor detectors at air intakes such as air conditioners are prudent (API, 1979). This section describes the types, response times, and positioning of remote vapor sensors and detectors.

5.1.1 Types of Sensors

There are several types of vapor detectors that operate on different principles (Chlorine Institute, 1980; Dailey, 1976; Harbert, 1984; Johanson, 1976; Krigman, 1984; Warncke, 1977; Wright and Wright, Inc., 1984; Zanetti, 1986).

5.1.1.1 COMBUSTION

The presence of flammable vapors can be detected by the heat developed from combustion on a hot wire. Typically, the hot wire is in an electrical bridge circuit to compensate for changes in ambient temperature. Since combustion cannot occur without an oxidizer, the atmosphere being sampled must have sufficient air to oxidize the sample; for example, analysis for a flammable hydrocarbon in a nitrogen atmosphere may give a zero concentration regardless of the hydrocarbon concentration. Such detectors cannot be made specific to particular gases or vapors; in fact, some vapors give readings that are incorrect by several percent, and correction factors need to be applied.

5.1.1.2 CATALYTIC

The reaction, combustion, or decomposition of some vapors during adsorption on catalysts can be used to determine concentration in air through heat effects (change in electrical resistance) or voltage developed. This method offers a quick response, but the catalysts may be poisoned by contaminants and the sensors generally cannot discriminate among similar materials.

5.1.1.3 ELECTRICAL

Special solid-state electrolytic sensors can change resistance when exposed to chemical vapors. Nonevaporating electrolytes also can be used to detect reduction reactions occurring upon exposure to some vapors. This type of sensor cannot discriminate very well among contaminants.

5.1.1.4 CHEMICAL REACTION

The chemical reaction between a vapor and a liquid adsorbed on a solid (e.g., silica gel) can result in a color change, produce a voltage, change the conductivity of a solution, or change the composition of a chemical compound. Color change is used in tube-type air samplers (grab samplers) and personal dosimeters (worn on the clothing). Some continuous analyzers employ electrolytic cells and produce a direct concentration reading. Analyzers that draw an air sample through a liquid for later analysis are sensitive to very low concentrations, but suffer from a long delay between sampling and results. Typically, chemical-reaction analyzers can be made specific to the desired vapor and can discriminate among contaminants.

5.1.1.5 VISUAL

The use of television cameras and monitors is limited to gases and vapors that are opaque, produce fog or aerosols (cryogenic or acid gases), or are strongly colored. So that the TV monitor does not have to be constantly attended, a system is available to sound an alarm if there is a sudden change in contrast in the picture being displayed.

5.1.1.6 ABSORPTION/SCATTERING

Several types of remote sensors use the property of gases and vapors to absorb or scatter specific wavelengths of infrared, ultraviolet, or other electromagnetic radiation (laser). By comparing the amount of light returned from a mirror or from the atmosphere at a specific wavelength with that of a standard unabsorbed wavelength having an equal path length, the concentration of the gas or vapor can be determined, together with its angular location from the light source. These systems may be affected by water vapor (fog) over long light-path distances and usually must be "tuned" to the specific material of interest.

5.1.2 Response Time of Sensors

The speed of response of a detector to a leak of an air contaminant may be extremely fast (milliseconds, for a remote sensor aimed at the point of leakage) to very slow (hours, for absorption in chemical solution for later analysis). For some fixed-point sensors, the delay may depend primarily on the distance from the leak source and the wind speed; for others, the delay may depend on the length of sample tubing between the pick-up point and the analyzer and the velocity of the sampled vapor through the tubing, which is typically about 10 feet per second (Lees, 1980, p. 534). For analyzers receiving inputs from several locations, the maximum cycle time should be determined by the distance to the nearest exposure (personnel or ignition source) divided by the typical wind speed, in consistent units. Table 5-1 compares several characteristics of different types of detectors, including sensitivity, selectivity of specificity, and response time.

TABLE 5.1 Evaluation of Gas Sensing Methods[a]

Principle	Parameter[b] Sensitivity	Parameter[b] Selectivity	Parameter[b] Speed of response	Sample system	Typical sensitivities[c] 0.01 ppm	Typical sensitivities[c] 0.1 ppm	Typical sensitivities[c] 1 ppm	Typical sensitivities[c] 10 ppm	Typical sensitivities[c] 100 ppm
Optical									
Infrared absorption	6	9	6	Yes/No				CO	$VC, SO_2, COCl_2$
Ultraviolet absorption	7	6	6	Yes/No			O_3, SO_2, H_2S		Cl_2, NO_2
Fluorescence	8	8	6	Yes		S			
Flame emission	8	6	6	Yes	P				
Chemiluminescence	8	8	6	Yes	O_3	NO_x			
Colorimetry	6	8	3	Yes		$COCl_2$	Cl_2, $HC, COCl_2$	SO_2, H_2S	
Laser	7	8	9	No					
Electrochemical									
Conductometry	6	4	6	Yes		HCl	SO_2, Cl_2		
Galvanometry	8	5	4	Yes		$H_2S, COCl_2$ SO_2, NO_2	HCN, NO_2		
Coulometry	6	6	4	Yes				CO	CO
Polarography	6	6	6	Yes			H_2S, NO_2	NO	
Potentiometry	6	4	8	Yes		F^-	$Cl^-, S=$		
Ionization									
Hydrogen flame	9	5	8	Yes			$=C=$		
Photoionization	8	2	8	Yes			$>10eV$		
Aerosol ion current absorption	8	4	6	Yes			HF, F_2, Cl_2	HCN	
Chromatographic									
Gas chromatograph	9	9	2	Yes			HC		
Thermal									
Conductivity	3	2	6	Yes				HC	
Catalytic combustion	8	7	9	No				HC	
Semiconductor	9	2	9	No			HC		
Human	6	8	3	No	H_2S	CS_2, Cl_2	$COCl_2, SO_2$	HCl, NH_3	$MeOH$

a References: Warncke (1977); Dailey (1976); Fazzalari (1978).
b Numbers represent ratings from very poor (0) to very good (10).
c VC = Vinyl chloride; HC = hydrocarbon.

5.1.3 Positioning of Sensors

The distance from a potential leak source to the sensors should be minimal for rapid response to leakage and notification of people in the affected areas. However, it may be more efficient to place sensors at a distance from a number of potential leak sources so that coverage of a larger area is provided when it is not possible to predict all potential leak sources. Ideally, fixed-point detectors should be placed around the perimeter of potential leak sources (Johanson, 1976; Lees, 1980, pp. 469, 534-536), with an angular separation that depends on atmospheric stability and wind speed (for example, about 10 degrees, or 36 sensors). Adjustments may be made for prevailing wind direction (Lihou, 1985) and the direction of prominent exposures so that the number of sensors could be reduced. The angular separation should remain at about 10 degrees and the separation between detectors should not exceed 30 feet (Johanson, 1976).

The recommended height for fixed-point detectors is 1.5 feet above the ground for vapors denser than air and 6 to 8 feet above the ground for light vapors and gases (Johanson, 1976).

Sensors need to be calibrated frequently because of their tendency to drift from zero, decreasing the sensitivity to alarm conditions or producing false alarms, and their tendency to decrease in sensitivity because of component deterioration or poisoning of reactive chemicals or catalysts. Records should be maintained to justify more frequent calibrations if unsatisfactory performance is observed, or less frequent calibrations if long periods of satisfactory performance are observed.

An insufficient number of detectors or poorly maintained detectors may be worse than none at all, because

- the number of personnel in the field or frequency of field patrols may be reduced as the detectors are installed, placing great reliance on the detectors; and
- action on field reports of leakage may be delayed by conflicting absence of control room alarms.

5.2 DETECTION BY PERSONNEL

Workers on plants handling hazardous materials can be trained to recognize potentially hazardous vapors by a specific or unusual odor or by visual observation. Although on-site personnel are not expected to be used as "detectors" in the same sense as in the previous section, they nevertheless need to be alert to the presence of vapors in the area. This may indicate a dangerous situation, and proper corrective action (such as leak/break isolation and area evacuation) may be required according to a predetermined response plan (as discussed in chapter 7).

5.2.1 Odor Warning Properties

Few chemical processes are operated entirely unattended, and personnel may be in the area of an accidental release. Many chemical substances have a characteristic odor or taste that would warn of a leak. However, some chemicals are odorless, such as carbon monoxide, dimethyl sulfate, natural gas, ethylene, hydrogen, and nitrogen. Table 5-2 compares odor thresholds for some commonly handled hazardous vapors with one measure of dangerous concentration (the IDLH level, as defined in Section 2.1).

Table 5-2. Comparison of Odor Thresholds with IDLHs

Gas or vapor	Odor threshold (ppm)	Immediate danger to life and health (ppm)	IDLH/odor
Carbon monoxide	(none)	1,500	0.000
Dimethyl sulfate	(none)	10	0.000
Chloroform	670	1,000	1.5
Ethylene oxide	500	800	1.5
Acrolein	0.25	0.5	2.0
Phosgene	0.47	2	4.3
Methyl isocyanate	4	20	5.0
Carbon tetrachloride	47	300	6.5
Ethyl ether	1,900	19,000	10
Hydrogen chloride	10	100	10
Hydrogen cyanide	2	50	25
Cyclohexane	400	10,000	25
Ammonia	21	500	25
Toulene diisocyanate	0.21	10	50
Chlorine	0.31	25	80
Formaldehyde	1	100	100
Aniline	1	100	100
Sulfur dioxide	0.47	100	200
Bromine	0.047	10	200
Methyl alcohol	100	25,000	250
Carbon disulfide	0.1	25	250
Chloroprene	0.8	400	500
Acetone	47	20,000	500
Benzene	2	2,000	1,000
Methyl methacrylate	0.21	4,000	20,000
Butadiene (1,3)	1.1	20,000	20,000
Nitrobenzene	0.005	200	40,000
Methyl mercaptan	0.001	400	400,000
Hydrogen sulfide	0.0002	300	1,500,000
Butyl mercaptan	0.0006	2,500	4,000,000

Odor may not provide adequate warning if it lessens or deadens the sense of smell. It is likely that all odors have a desensitizing effect that varies with time. Hydrogen sulfide causes olfactory fatigue in less than 1 minute (Cheremisinoff and Young, 1975; Eugen, 1982) while chlorine takes 10 minutes (NIOSH, 1976).

Some individuals cannot detect the odor of certain gases and vapors because of genetic traits or overexposure to irritants that cause loss of smell (Clayton and Clayton, 1978, 1982). It may be necessary to test personnel to identify people who are unable to detect a contaminated atmosphere by smell. There are a few materials, such as the mercaptans, that have odor thresholds well below the IDLH concentrations. Odor should serve as a dual warning that leakage is occurring and needs correction and that the area should be evacuated until breathing protection is obtained (Lees, 1980, pp. 645 and 838). Personnel should not remain in an area where there is an odor because people are not able to calibrate odor against concentration very well, particularly with the present effort to minimize the size, duration, and frequency of any chemical emissions.

5.2.2 Color or Fog

Few gases and vapors have characteristic colors that can be used to identify leaking material and to estimate their concentration. One colored gas is chlorine, which is greenish-yellow, but little information is available concerning the relationship between its color and concentration. *The Chlorine Manual* (Chlorine Institute, 1986, p. 5) states only that "its greenish yellow color makes it visible at high concentrations." Other gases and vapors which have noticeable color are nitrogen dioxide (reddish brown), bromine (red-orange), and chlorine dioxide (greenish yellow to orange). It is likely that these materials are detectable by color only at concentrations above 1 percent (10,000 parts per million) and provide directional and concentration warnings only for personnel close to the leakage source (and then only during daylight hours or with very strong illumination).

Many vapors become visible when mixed with humid air by extracting moisture from the air to form aerosols. Examples include hydrogen chloride, hydrogen fluoride, sulfur trioxide, phosphorous trichloride, phosphorous oxychloride, silicon tetrachloride, titanium tetrachloride, antimony trichloride, antimony pentachloride, and ammonia. Vapors from liquefied gases at temperatures well below the existing dew point can cause fog by condensing moisture from humid air. Examples include propane, methane, ethylene, and ammonia. Thus, the aerosol-producing characteristics of some toxic and flammable materials can be expected to indicate the source of leakage and to provide indications of wind direction and speed. Closed circuit television can be used to monitor large areas of a plant for aerosol-indicated leakage.

As a result of recent tests (Du Pont, 1983), a correlation of vapor concentration, relative humidity, heat of solution, and vapor or aerosol visibility can be derived (Appendix C). However, it is not likely that concentration and release rate information obtained from visible cloud dimensions could be used in the short time available between release and exposure.

Visual detection of colorless, foglike vapors and aerosols can be hindered by the presence of fog. Vapor detectors (other than ultraviolet, infrared, or laser systems) are preferable to visual observation in detecting the occurrence and extent of leakage if the probability of fog is high.

5.3 ALARM SYSTEMS

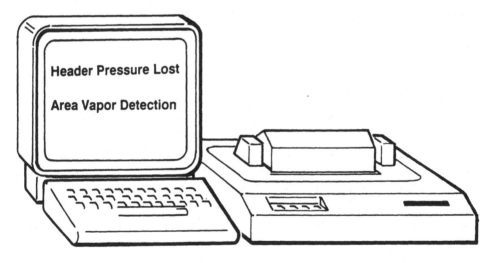

Air contaminant or leak detectors transmit an alarm signal to an attended area such as a control room, security office, or gate house. The attendant must be trained to respond to the alarm and to take the following actions nearly simultaneously:

- Sound alarms to alert personnel in the immediate area of the release to determine the source of the leak, attempt to stop the leak, evaluate the size of the leak, shut down processes, or evacuate the area, depending on the emergency response plan.
- Notify people downwind of the leak or release by alarms, telephone, or radio. The extent of the area to be warned will de-

pend on the severity of the release and the degree of toxicity/flammability.
- Notify management--important decisions may be needed regarding long-range evacuation, notification of off-site authorities, and further implementation of emergency-response or disaster plans. Emergency shutdown procedures and emergency response communications are discussed in chapters 7 and 8.

Television-like display systems are available to present vapor detector positions within building or operating areas. Alarm symbols appear and change (by flashing or changing color) if preset concentrations of gas or vapor are exceeded. Alarm messages regarding the position and concentration can be displayed. With presentation of wind direction, the source of leakage can be approximately located and the leak size estimated. Computerized alerting systems which do not use vapor detector input, but predict the vapor plume after a leak size is estimated, are discussed in Section 7.1.

Many types of computer aids are available to streamline emergency response and reduce delays. Process computers can be programmed to detect leakage from changes in pressure, flows, and levels to support, verify, or combine with alarms generated by leak detectors. Once a reliable system is established and false alarms decrease, area alarms, site-wide alarms, and off-site warnings can be automated.

REFERENCES

API, 1979: *Safety Digest of Lessons Learned*, Publication 758, Section 2, American Petroleum Institute, New York, 157.

Cheremisinoff, P. N., and R. A. Young, 1975: *Industrial Odor Technology and Assessment*, Ann Arbor Science Publishers, Ann Arbor, MI, 20 and 53.

Chlorine Institute, 1980: *Atmospheric Monitoring Equipment for Chlorine*, 3rd Ed., The Chlorine Institute, Washington, D.C.

Chlorine Institute, 1986: *The Chlorine Manual*, 5th Ed., The Chlorine Institute, Washington, D.C.

Clayton, G. D., and F. E. Clayton, 1982: *Patty's Industrial Hygiene and Toxicology*, Vol. IIc, Interscience and Wiley, New York, 4853.

Clayton, G. D., and F. E. Clayton, 1978: *Patty's Industrial Hygiene and Toxicology*, Vol. I, Interscience and Wiley, New York, 674.

Dailey, W.V., 1976: Area monitoring for flammable and toxic hazards. *Loss Prev. 10,* 8.

Du Pont, 1983: Private communication, E. I. du Pont de Nemours and Company, March 14 and August 23.

Eugen, T., 1982: *The Perception of Odors*, Academic Press, New York, 64.

Fazzalari, F. A. (ed.), 1978: *Compilation of Odor and Taste Threshold Value Data*, Data Series DS-48A, American Society for Testing and Materials, Philadelphia.

Harbert, F., 1984: Detection of fugitive emissions using laser beams. *The Chem. Eng.*, (407), 41.

Johanson, K. A., 1976: Design of a gas monitoring system. *Loss Prev. 10*, 15.

Kletz, T. A., 1975: Emergency isolation valves for chemical plants. *Chem. Eng. Prog. 71* (9), 73; *Loss Prev. 9*, 134.

Krigman, A., 1984: Toxic gas monitoring: Simple instruments--complex measurements. *Instrum. Technol. 31* (5), 487.

Lees, F. P., 1980: *Loss Prevention in the Process Industries*, Butterworth, London.

Lihou, D., 1985: Why did Bhopal ever happen? *The Chem. Eng.*, (413), 18.

NIOSH, 1976: *Criteria for a Recommended Standard: Occupational Exposure to Chlorine*, Publication No. 76-170, National Institute for Occupational Safety and Health, U.S. Govt. Printing Office, Washington, D.C., 32.

Warncke, H., 1977: Monitoring systems for toxic gases. *Int. Symp. Loss Prev. 2nd (Heidelberg)*, 487.

Wright and Wright, Inc., 1984: Monitor spots hydrocarbon leaks. *Chem. Eng. 91* (21), 49.

Zanetti, R., 1986: Remote sensors zero in on toxic gas targets. *Chem. Eng. 93* (5), 14.

6

MITIGATION THROUGH COUNTERMEASURES

Toxic or flammable liquids, liquefied vapors, vapors, or gases that cannot be prevented from escaping from the process system by isolation valves, transfer systems, or shutdown switches must be combated by some other method. Several methods of counteracting releases are available. These capabilities should be considered in the design of a process and be included in plans for coping with accidental releases. Many of them use a reliable fire-fighting water supply. Some methods require manual (active) operation of equipment. Their effectiveness depends on the training and number of personnel on the site at the time a release occurs, as well as on the frequency of drills.

6.1 VAPOR/LIQUID RELEASES

The countermeasures employed in a given situation will be different depending upon the nature of the release being combated. A vapor cloud already formed and beginning to drift downwind will have one set of potential countermeasures. A liquid spill may have an altogether different set of potential countermeasures. In the following sections, a "liquid spill" implies a release of volatile liquid which may or may not be boiling after the spill occurs.

In many cases, countermeasures will need to be taken against both the vapor cloud and the liquid spill. This is especially true when dealing with a flashing liquid, as described in Section 1.4, where a vapor cloud is generated and a boiling/evaporating liquid pool is formed simultaneously. Such a case may be further complicated by aerosol droplets in the vapor cloud.

6.2 VAPOR RELEASE COUNTERMEASURES

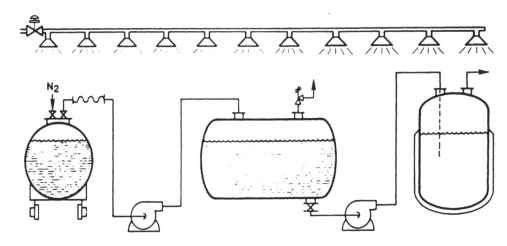

Water, steam, and air curtains and water sprays are primarily effective in dispersing and/or diluting vapors with air to reduce the severity of effects of a hazardous vapor release. In some cases, vapors can be partially "knocked down" or absorbed after release. Ignition source control or deliberate ignition are also possible vapor release counter-measures.

6.2.1 *Water Sprays*

Water in the form of "fog" can be used to partially absorb water-solu-ble gas or vapors (Lees, 1980, p. 470). A fine spray at high flow rates, upwind and towards the release source, is needed. However, water sprays will knock down only a fraction of the vapors. Other limita-tions to consider are that the water sprays must be carefully posi-tioned to be effective, and the vapors may travel downwind and out of range of the water sprays before they can be employed.

Water sprays can dilute vapor clouds by mixing the vapors with the air entrained by the spray of water droplets (Beresford, 1981; Smith, 1984). Hand-held nozzles can sometimes maneuver vapor clouds away from ignition sources if the vapor cloud is visible because of aerosol formation or condensation of moisture in the air. Of course, such intervention is accompanied by a risk of exposure to toxic contaminants or thermal radiation if a flammable vapor cloud ignites, and appropriate protective clothing and concentration monitoring is required.

Fixed water spray systems can partially suppress the release of noxious soluble vapors from open pits that receive spills through trenches. The spray pattern should obscure the surface, but the spray should not be so dense that vapors are pumped out of the pit by air entrained with the spray.

Experience. Ammonia vapor clouds have been controlled and absorbed by four 1.5-inch or 1.75-inch fog nozzles positioned downwind of spills and aimed toward the spill (Greiner et al., 1984). The optimum location for the nozzles was about 100 feet downwind of the 5-pounds-per-second release for a 5-mph wind (with a 45-degree pattern) to 50 feet for winds greater than 15 mph (with a 30-degree pattern). Based on a flow rate near 100 gallons per minute per nozzle and a 30-foot separation of the nozzles, the spray density would be about 0.2 gallons per minute per square foot. The ammonia concentration at a distance of 200 feet was reduced to about 200 ppm from a calculated centerline concentration of 15,000 ppm (Greiner, 1987).

6.2.2 Water Curtains

A curtain of water may be used to separate a leak from personnel and sources of ignition. A water curtain is a spray of water from a horizontal pipe through nozzles. The curtain may be actuated manually or automatically in response to vapor detectors or to reports by personnel in the area (visual, audible, or odor indications of leakage).

The water curtain dilutes vapors with entrained air (McQuaid and Fitzpatrick, 1981; Zalosh, 1981; Moodie, 1984). The water spray also can absorb vapors if they are highly soluble, as are many acid vapors; however, total absorbtion should not be expected. Water curtains are only marginally effective in reducing the hazard of vapors not soluble in water (Deaves, 1983; Harris, 1980; Lees, 1980, pp. 470, 671; McQuaid, 1977; Vincent et al., 1976; Emblem and Madsen, 1986). The vapor cloud may be diluted but the effects of the curtain disappear (Meroney et al., 1984) when the vapor is at a distance from the leak source, since

- the vapor is dispersed rather than absorbed, and the weight rate of vapor flow downwind is essentially unchanged, particularly for dense vapors; and
- the air entrained by the water sprays eventually becomes contaminated with the vapor being dispersed (by vapor flow over and/or around the curtain), particularly for dense vapors.

Turbulence is generated by mixing the vapor and entrained air. If the vapor downstream of the curtain is above the lower flammable limit when the cloud reaches an ignition source, the flame can pass through the curtain, and turbulence induced by the curtain may intensify the combustion (Lees, 1980, pp. 470, 671). Also, the concentration of flammable or toxic vapor may be increased upwind of the curtain (Emblem and Madsen, 1986).

Although no tests are reported on materials that are very soluble in water, it is expected that the concentrations downstream of the curtain may be much lower than those at the same downwind position without the water curtain. In a study of hydrogen fluoride hazards (Health and Safety Executive, 1978), water sprays were recommended to absorb escaping vapor. An effectiveness of 80 to 90 percent was assumed, but no design information was given. Installation of water curtains in regions having strong winds may be inadvisable.

Design. The quantity of air that must be entrained by the water curtain to effect the required dilution must be determined in designing dilution-effect water sprays (Vincent et al., 1976). Based on an experimentally derived ratio of the entrained air rate to the water spray rate, a required water flow is calculated.

For essentially insoluble vapors, such as hydrocarbons, a decrease in concentration by a factor of about four could be expected with a well-designed water curtain having (1) a length equal to twice the distance from the leak source and (2) a water spray rate about five times greater than the release rate (for example, 350 gallons per minute for a 10-pounds-per-second leak). For very soluble vapors, such as hydrogen chloride and hydrogen fluoride, an order-of-magnitude reduction in concentration could be expected with the same length and water spray rate. Water curtains have been installed to isolate sources of ethylene leakage from ignition sources (Meroney et al., 1984).

Water curtains have been installed for protection of the public from hydrogen fluoride vapors at two locations in England (Health and Safety Executive, 1981). At another location, a double ring of water sprays around a chlorine storage sphere forms a solid sheet of water and provides partial protection against leakage from the tank or bottom connections.

6.2.3 Steam Curtains

Steam curtains operate similarly to water curtains except that the absorbing effect is minimal. The higher fluid temperature enhances the buoyancy of vapors passing through the curtain (Moore and Rees, 1981; Rulkens, 1984). Cold vapors with a molecular weight below 29 become naturally buoyant and can rise above downwind exposures and ignition sources, while those with a specific gravity above 1.0 become buoyant (Cairney and Cude, 1971; Lees, 1980, p. 469) long enough to mix with air and attain neutral buoyancy.

Typically (Cairney and Cude, 1971; Lees, 1980, p. 469; Barker et al., 1977), a horizontal steam pipe with a row of small holes is mounted near the top of a 4- to 5-foot wall. A 6-inch-diameter pipe with a row of 5/32-inch holes spaced 4 inches apart, with 250-psig steam supply pressure, reduced concentrations by a factor up to 30 with equal steam and contaminant flow rates. Jets from the holes combine to form a planar curtain that entrains air to dilute the vapor cloud below its lower flammability limit. Sections in the curtain piping are individually controlled by valves (Bockmann et al., 1981) to allow steam to be directed in the most effective pattern.

Steam requirements for steam curtains are considerable. To reduce concentrations below the lower flammable limit, about 0.2 pound of steam per pound of contaminant is required (Cairney and Cude, 1971; Lees, 1980, p. 469), and these quantities may be impractical. Supply and distribution systems would be practical only for limited areas (such as pump alleys). A steam curtain requires about 100 pounds per hour of 250-psig steam per foot of curtain, which is about 80 times the energy requirement per foot as a water curtain (McQuaid, 1977).

Tests of static-electricity generation by steam curtains (Seifert et al., 1983) indicate that ignition of hydrocarbon/air mixture is unlikely if all components are grounded.

6.2.4 Air Curtains

Few tests have been made with air curtains (Rulkens et al., 1983). The air supply is usually limited at most plants and air curtains would not provide the vapor heating of steam curtains.

6.2.5 Deliberate Ignition

Deliberate ignition is a countermeasure against spills of highly toxic materials which are also flammable, such as hydrogen sulfide, hydrogen cyanide, and methyl mercaptan. Igniting nontoxic flammable materials such as hydrocarbons may present hazards outweighing possible advantages. The objective of deliberate ignition of a toxic liquid or vapor release is to destroy the toxic material before the vapor can travel to people. Also, making the vapor cloud more buoyant may reduce ground-level concentrations. Note, however, that the vapors may be dangerously toxic well below the lower flammable limit.

In procedures for deliberate ignition, the potential for serious aggravation of the emergency must be considered, such as the spreading of a fire or the initiating of a vapor cloud explosion (Bodurtha, 1980, pp. 102, 110; Welker, 1969). Delayed ignition makes aggravation of the situation more likely. Even if ignition is accomplished, downwind evacuation may still be required because of incomplete combustion of the toxic vapors (Husa and Bulkley, 1965) and/or combustion products which are also toxic.

Remote ignition devices include flare pistols, electric ignitors, and propane burners. The pistols should be deployed in four quadrants around potential spillage areas, for accessibility regardless of wind direction. Electric ignitors (such as spark plugs) and propane burners should be deployed at about 3.6 degree intervals around the plant with each about 4 feet above ground (Lee et al., 1977). They should be in constant operation or activated by flammable vapor detectors to preclude formation of a large cloud.

6.2.6 Ignition Source Control

For areas around processes handling flammable vapors, ignition source control is practiced to reduce the probability of vapor ignition if a leak occurs. Many incidents have occurred in which large quantities of flammable vapors were accidentally released but no fire or explosion ensued, because the vapors dispersed without igniting.

Ignition source control must attempt to control all likely ignition sources in a sufficiently broad area around the process. Potential ignition sources include

- furnaces, kilns, incinerators, and boilers;
- operating flares;
- hot work by maintenance or construction activities (e.g., welding);
- smoking or use of matches;
- sparking electrical equipment;
- vehicle operation;
- hot surfaces;
- electrostatic discharges, including lightning;
- mechanical/friction heat generation; and
- an external accidental fire, which may have arisen from the same cause as the vapor release.

No attempt will be made in these guidelines to cover control of all ignition sources. A few pertinent considerations will be addressed.

The National Electric Code and National Fire Codes address various area classifications with respect to electrical equipment; however, it should be noted that limited-extent area classifications are only intended to protect against minor, occasional leaks where the extent of a flammable concentration is 50 feet or less (NFPA, 1986a). Major, accidental flammable vapor releases can be expected to travel far beyond normal area boundaries and may easily reach electrically unclassified areas. Where hand-held or transportable electrical devices are considered for use, the likelihood of their being a potential ignition source should be assessed. Buildings in which "general purpose" electrical equipment is installed should have the capability of prompt isolation from the outside environment by closing air intakes and shutting off exhaust ventilation systems (including laboratory hoods and restroom fans).

Hot surfaces may include process equipment as well as electrical and mechanical equipment. Specially designed equipment has been developed for some chemical-specific applications, such as motors designed to operate at surface temperatures below the very low autoignition temperature of carbon disulfide (100°C).

Administrative controls are exercised on plants where flammable materials are processed. Such controls may include hot work permits, restricted smoking areas, not allowing lighters or matches on the site, and electrical grounding and bonding procedures. Some areas may warrant such measures as humidity control and the wearing of static-control clothing and footwear, use of nonsparking tools, and special forklift trucks which minimize ignition potential. Shutdown of units such as furnaces which may have open flames usually requires careful analysis of, for example, the times required to detect the vapor release, communicate a shutdown directive to the area, and allow the unit to cool down sufficiently, as compared to the time required for the vapor cloud to travel from the process to the furnace.

6.3 LIQUID RELEASE COUNTERMEASURES

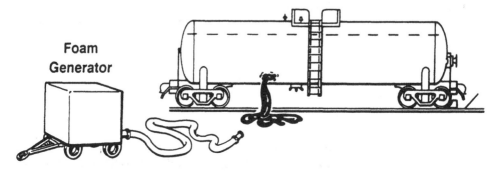

Foam
Generator

Practical methods for combating vapor from liquid leaks are dilution, neutralization, or covering. All three reduce the vaporization rate of the pool.

6.3.1 Dilution

Water dilution is effective for spills of water-miscible or water-soluble material. Spraying water into the spill reduces the vapor pressure by reducing the concentration of the liquid (Lees, 1980, pp. 471, 671). Although usually not appropriate, other liquids (such as high-flash-point oils) can be used for small spills. When the heat of solution is high, adding water may increase the vaporization rate, thus requiring a large volume of rapidly applied water. The amount needed can be estimated from an enthalpy diagram.

Emergency plans for most reactive liquids and liquefied gases should not be based on dilution. Water aggravates the hazard for many liquefied gases (such as chlorine) by providing heat for vaporization, for materials that react with water by creating additional hazards (such as acetone cyanhydrin, which releases hydrogen cyanide on contact with water), and for those that react violently with water (such as phosphorous, many hydrides and silanes, and methyl isocyanate). For these, alternative countermeasures such as covering spills with nonreactive barriers are recommended.

6.3.2 Neutralization

An appropriate response to a spill emergency can be reacting the spilled liquid with a material to achieve a less-volatile salt or ester. However, the neutralizing agent must be delivered to the spill at a high rate, much greater than the stoichiometric quantity, to avoid aggravating the vapor release hazard. For acidic spills, limestone or soda ash is often used; caustic soda or solutions create a corrosivity hazard.

Fire-extinguishing equipment has been used to apply neutralizing and solidifying agents to spills of acid or caustic material from a distance (Ansul, 1986). A typical ratio is a pound of agent to a half pound of spilled acid or base, which results in a neutral solid that can be disposed in a landfill.

6.3.3 Covers

6.3.3.1 LIQUIDS
Dense liquids can be covered with lighter immiscible nonreactive liquids to prevent vaporization. Bromine (specific gravity 2.9, boiling

point 59°C) can be covered with water (in which it is only slightly soluble), preferably with sodium bicarbonate added to further reduce vaporization. A water cover protects a spill of carbon disulfide (specific gravity 1.3, boiling point 46°C) from ignition and evaporation. Chlorosulfonic acid (specific gravity 1.77) can be covered by a less dense aliphatic oil to prevent fuming until neutralization and disposal can be accomplished. A couple of inches of liquid covering the entire surface is sufficient to prevent vaporization in most situations.

6.3.3.2 FOAMS

A foam cover can be effective in reducing vaporization from spills (Lees, 1980, pp. 268, 427, 471, 773). Foam covers protect spills from heating by solar radiation and the atmosphere; they also act as a barrier to prevent mass transfer from the liquid to the air. Various chemically resistant foams are available for hydrocarbon, acid, or alkaline spills (Hiltz, 1982; Norman and Dowell, 1980; Nat. Foam, 1986; Jeulink, 1983; Welker et al., 1973, 1974; DiMaio and Norman, 1987). Such foams are effective against the hazard of pooled spills, particularly if the spills are confined within dikes, curbs, or trenches, but not against running spills or those that are on fire.

Fire protection foam makers (Nat. Foam, 1986) have been used to proportion the foam and water solution at a ratio of 6 gallons of foam to 100 gallons of water, producing an expansion ratio of 15:1 and a 25 percent drainage time of 12 minutes or more. Four inches of foam can reduce vaporization rates by 40 percent or more. Cannons or monitors can apply foam remotely (Jeulink, 1983). Engineered foams (Nat. Foam, 1986) are intended to be compatible with the material to be covered, but there may be reactions with the water in the foam; thus, aqueous foams are not recommended for water-reactive spills (e.g., chlorosulfonic or fluorosulfonic acids, anhydrous hydrogen fluoride, phosphorous or sulfur chlorides, chlorosilanes, phosgene, or hydrogen cyanide).

Foam systems can be engineered similar to fire protection systems to suppress vaporization of toxic or flammable liquids in sumps, pits, or dikes (NFPA, 1986b, pp. *19*-32, *19*-41, *19*-43). The foam blanket should be about 4 inches thick for fire-fighting foam with an 8:1 expansion ratio, or about 6 inches thick for special foams with a 20:1 expansion ratio (Nat. Foam, 1986). For a typical application rate of 0.16 gallon per minute per square foot (NFPA, 1983), the application would only take about 2 minutes, provided that foam destruction (by the liquid being covered) does not occur. The amount of foam concentrate required, at a 6 percent concentration, would then be about 0.02 gallon per square foot of dike, sump, or pit area.

6.3.3.3 SOLIDS

Granular materials can reduce vaporization by effectively reducing the exposed area. They are usually less dense than the spill, not reactive with it, fine enough to settle on it, and can be shoveled or blown onto the spill. Sand, floating balls (Harris, 1980), polymer beads, and sulfate powders (ICI, 1980) have been used on strong acids. Shallow spills can be "wicked" up, thereby reducing the vapor pressure (NFPA, 1986b, pp. *19*-32, *19*-41, *19*-43), or absorbing the spilled material (NFPA, 1986a; ICI Ltd., 1980). Plastic films have been tested for preventing vaporization from pools (Harris, 1980).

6.3.3.4 APPLICATION

Adding any material to a subcooled pool of liquefied gas will cause a temporary surge in vaporization rate. If the covering material has a low heat load (heat capacity times density), the surge will be brief and followed by a long-term lower vaporization rate. Table 6-1 lists specific covering materials that have been proposed or used.

6.4 AVOIDANCE OF FACTORS THAT AGGRAVATE VAPORIZATION

Some measures can be taken to avoid aggravating vaporization from spilled fluids, particularly liquefied gases. For example, there should be no loose gravel under tanks or process vessels containing liquids such as chlorine which have boiling points below ambient temperature (Harris, 1980). The gravel can serve as an instantaneous heat source for the boiling liquids (Harano, 1976) and cause ground-contact flashing. This may add considerable vapor to clouds already generated by adiabatic flashing.

Further, the ground under storage tanks should not be reactive with the material stored in the tanks unless it is likely that the net result of a reaction is reduced hazard. For example, silica materials (such as sand) should be removed from under tanks of anhydrous hydrogen fluoride because reaction would produce heat and toxic silicon tetrafluoride. However, limestone under hydrogen fluoride tanks may result in lowering the hazard of leaks because calcium fluoride would be formed, even though considerable heat may be evolved. Concrete generally is not a suitable material for use under acid storage tanks (because of the heat of reaction), and asphaltic materials are not suitable under tanks of oxygen (Lees, 1980, p. 268) or chlorine (because a fire may ignite or chemically reactive products may be formed if a leak occurs).

Table 6-1. Materials Used to Cover Vaporizing Liquid Spills

Spilled material	Covering material	Reference[a]
Acetic anhydride	Oil, 1/4"; vermiculite, 2"	Small, 1974
	Tridecanol, 1/4"	Small, 1974
Acrolein	7% Sodium carbonate	Small, 1974
Acrylonitrile	Polar-liquid foam	Norman, 1980
	Activated carbon, 2"	Small, 1974
	Carbonaceous powder	Ansul, 1986
Amines	Activated carbon, 2"	Small, 1974
	Citric-acid-based powder	Ansul, 1986
Ammonia and amines	Alkali-resistant foam	Nat. Foam, 1986
Bromine	Acid-resistant foam	Nat. Foam, 1986
	Water	
Butane	Polar-liquid foam	Norman, 1980
Carbon disulfide	Protein foam	Norman, 1980
	20% Sodium hydroxide; water	Small, 1974
	Water	*
Chlorine	Water-based protein foam	Lees, 1980, p. 671
	Acid-resistant foam	Nat. Foam, 1986
Chlorosulfonic acid	Polyacrylamide 1916 powder	ICI, 1980
Ethylene oxide	Polar-liquid foam	Norman, 1980
Gasoline	Polypropylene beads	Welker, 1973
	Carbonaceous powder	Ansul, 1986
Hydrogen chloride	Acid-resistant foam	Nat. Foam, 1986
Hydrogen fluoride	Polyacrylamide powder	ICI, 1980
	Oil, paraffin-base, 1/2"	Small, 1974
	Magnesium-oxide powder	Ansul, 1986
Liquid hydrocarbons	Alcohol-resistant fluorocarbon surfactant foam	Jeulink, 1983
	Carbonaceous powder	Ansul, 1986
Liquefied natural gas	High-expansion foam	Norman, 1980
Oleum (65%)	Granular sodium sulfate	ICI, 1980
	Magnesium oxide powder	ICI, 1980
Phosgene	Kerosene and dry sawdust	Lees, 1980, p. 671
Phosphorus trichloride	Sand	*
Silicon tetrachloride	High-expansion foam	Hiltz, 1982
Sulfur trioxide	Polycarbonate beads	ICI, 1980
Titanium tetrachloride	Acid-resistant foam	Nat. Foam, 1986

[a]* = industrial experience (no reference).

REFERENCES

Ansul, 1986: *Spill Control Treatment Guide*; video tape, *Spill-X Spill-gun Applicators-- Application and Operation.* Ansul Fire Protection Company, Marinette, WI.

Barker, G. F., et al., 1977: Olefin plant safety during the last 15 years. *Loss Prev. 11,* 4.

Beresford, T. C., 1981: The use of water spray monitors and fan sprays for dispersing gas leakages. Inst. Chem. Eng., North West. Branch Papers No. 5, 6.1.

Bockmann, T., et al., 1981: Safety design of the ethylene plant. *Loss Prev. 14,* 127.

Bodurtha, F. T., 1980: *Industrial Explosion Prevention and Protection*, McGraw-Hill, New York.

Cairney, E. M., and A. L. Cude, 1971: The safe dispersal of large clouds of flammable heavy vapors. *Inst. Chem. Eng. Symp. Ser. 34*, 163.

Deaves, D. M., 1983: Experimental and computational assessment of full-scale water spray barriers for dispersing dense gasses. *Int. Loss Prev. Symp. 4th (Harrogate)*, F-45.

DiMaio, L. R., and E. C. Norman, 1987: Performance of aqueous Hazmat foams on selected hazardous materials--A report. Paper presented at 1987 American Institute of Chemical Engineers Loss Prevention Symposium, Minneapolis.

Emblem, K. and O. K. Madsen, 1986: Full scale test of a water curtain in operation. *Int. Loss Prev. Symp. 5th (Cannes, France)*, 46-1.

Greiner, M. L., et al., 1984: Emergency response procedures for anhydrous ammonia vapor release. *Plant/Oper. Prog. 3* (2), 66.

Greiner, M. L., 1987: Telephone communication, March 16.

Harano, R., 1976: A report on the experimental results of explosion and fires of liquid ethylene facilities. Japanese Ministry of International Trade and Industry (MITI), Safety Information Center, Tokyo (English trans. Insurance Tech. Bur., London), 25.

Harris, N. C., 1980: The control of vapour emission from liquified gas spillages, *Int. Loss Prev. Symp. 3rd (Basel, Switzerland)*, 1058, 1061.

Health and Safety Executive, 1978: *Canvey: An Investigation of Potential Hazards from Operations in the Canvey Island/Thurrock Area*, HM Stationery Office, London, 18, 27, 30, 92, 98, 99, and 100.

Health and Safety Executive, 1981: *Canvey: A Second Report*, HM Stationery Office, London, 6 (correction slip), 56 and 58.

Hiltz, R. M., 1982: Mitigation of the vapor hazard from silicon tetrachloride using water-based foam. *J. Hazard. Mater. 5* (3), 169.

Husa, H. W., and W. L. Bulkley, 1965: Hazards of liquid ammonia spills. *Safety in Air and Ammonia Plants 7*, 42.

ICI, 1980: *Treatment of Fuming Acid Spillage*, motion picture, ICI Ltd., Porton Down, U.K.

Jeulink, J., 1983: Mitigation of the evaporation of liquids by fire-fighting foams. *Int. Symp. Loss Prev. 4th (Harrogate)*; *Inst. Chem. Eng. Symp. Ser. 80*, E12.

Lee, J. H., et al., 1977: *Dispersion of Gas Releases*, McGill University report, May 23.

Lees, F. P., 1980: *Loss Prevention in the Process Industries*, Butterworth, London.

McQuaid, J., 1977: The design of water-spray barriers for chemical plants. *Int. Loss Prev. Symp. 2nd (Heidelberg)*, 511.

McQuaid, J., and R. D. Fitzpatrick, 1981: The uses and limitations of water-spray barriers. Inst. Chem. Eng., North West. Branch Papers No. 5, 1.1.

Meroney, R. N., et al., 1984: Wind-tunnel simulation of field dispersion tests (by the U.K. Health and Safety Executive) of water-spray curtains. *Boundary Layer Meteorology 28* (1), 107.

Moodie, K., 1984: An experimental assessment of water spray barriers for dispersing clouds of heavy gases. *Inst. Chem. Engrs. European Branch Symp. Ser. 3 (Utrecht)*, 74.

Moore, P. A. C., and W. D. Rees, 1981: Forced dispersion of gases by water and steam. Inst. Chem. Eng., North West. Branch Papers No. 5, 4.1.

NFPA, 1983: Foam extinguishing systems. *National Fire Code*, NFC 11, National Fire Protection Association, Boston.

NFPA, 1986a: Classification of Class I hazardous locations for electrical installations in chemical process areas. *National Fire Code*, NFC 497A, National Fire Protection Association, Boston.

NFPA, 1986b: *Fire Protection Handbook*, National Fire Protection Association, Boston.

Nat. Foam, 1986: Controlling hazardous vapors. *Technical Manual*, Section XIV, National Foam System, Inc., Lionville, PA.

Norman, E. C., and H. A. Dowell, 1980: Using aqueous foams to lessen vaporization from hazardous chemical spills. *Loss Prev. 13*, 27.

Rulkens, P. F. M., et al., 1983: The application of gas curtains for diluting flammable gas clouds to prevent their ignition. *Int. Symp. Loss Prev. 4th (Harrogate)*; *Inst. Chem. Eng. Symp. Ser. 80*, F-15.

Rulkens, P. F. M., 1984: The application of steam curtains for the dilution of gas clouds. *Inst. Chem. Engrs. European Branch Symp. Ser. 3 (Utrecht)*, 105.

Seifert, H., et al., 1983: Steam curtains--effectiveness and electrostatic hazards. *Int. Symp. Loss Prev. 4th (Harrogate)*; *Inst. Chem. Eng. Symp. Ser. 80*, F-1.

Small, R. H., and G. E. Snyder, 1974: Controlling in-plant spills. *Loss Prev. 8*, 24.

Smith, J. M., 1984: The use of upward directed water sprays to disperse heavy gas clouds. *Inst. Chem. Engrs. European Branch Symp. Ser. 3 (Utrecht)*, 119.

Vincent, G. C., et al., 1976: Prevention of explosion by water fog. *Loss Prev. 10*, 55.

Welker, J. R., et al., 1969: LNG spills: To burn or not to burn. *Proc. 1969 Operating Section Conference*, American Gas Association, D-262.

Welker, J. R., et al., 1973: *Hydrocarbon Process. 52* (5), 105.

Welker, J. R., et al., 1974: Use foam to disperse LNG vapors. *Hydrocarbon Process. 53* (2), 119.

Zalosh, R., 1981: Dispersal of L.N.G. vapour clouds with water spray curtains. Inst. Chem. Eng., North West. Branch Papers No. 5, 3.1.

7

ON-SITE EMERGENCY RESPONSE

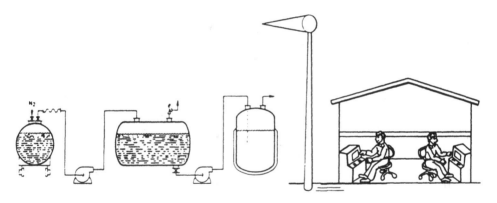

Protection of persons from injury, both on site and off site, is the primary objective of vapor release mitigation. In addition to countermeasures which attempt to control the vapor release from spilled material at its source (as described in the previous chapter), reducing the severity of the vapor release consequences can also be accomplished by protecting persons from hazardous vapors which do escape from the immediate area of the release. This chapter describes on-site responses to such a vapor release emergency. In addition to being quickly and effectively warned of a dangerous situation, personnel need to know ahead of time the best response to minimize their chances of being affected by the vapors. Questions need to be settled in advance, such as whether the process must be shut down in an emergency, whether it is better to stay indoors or attempt evacuation, and what respiratory protection will be effective in a given situation. In general, staying indoors in a temporary "haven" is better than trying to escape from a vapor cloud, if the person cannot escape without going through a hazardous area. This is particularly true for flammable vapor clouds, where the vapors may ignite while a person is within the flammable cloud. For massive flammable vapor clouds, blast effects of a possible vapor cloud explosion need to be considered as well.

7.1 ON-SITE COMMUNICATIONS

Communicating warnings to potentially exposed personnel is essential in an emergency. Such warnings should be initiated as soon as leakage is detected and an assessment of the hazard can be made, as discussed in Section 5.3. The authorization for initiating warnings and sounding evacuation alarms should be established at the lowest practical level in the organization and be clearly defined in the emergency plans. Otherwise, valuable time may be lost in determining if approval for warning is needed and, if so, in contacting the required management representative.

The warnings, to be properly communicated or understood (for example, if site-wide hazardous vapor or fume alarms are used), require considerable preplanning and training. As an example, warnings of toxic gas releases may include orders for evacuation or taking refuge in havens. The criteria for choosing the proper response should be stated in the emergency plans.

Hazard warnings are communicated in various ways. Most fire-alarm systems can be adapted to serve as vapor or "fume" alarms, with coded signals to indicate the location of the leak. It is important for all areas of the site to have an adequate number of alarm signals, particularly where noise may nullify a plant-wide alarm. At some sites, the sounding of a hazardous vapor or fume alarm is followed by radio messages that provide as much information on the release as is currently available.

Automatic telephone dialing and alerting ("seizure") systems are also used for communicating simultaneous hazard warnings to downwind personnel. These systems may transmit recorded messages to as many as 100 telephones in selected areas of a site. The telephones ring until answered, then signal the message transmitter that the call was completed. Computers can be programmed to dial preselected telephones through a modem. Such a system avoids stress-caused misdialings and greatly reduces the time required for telephoning warning messages. Self-dialing telephone equipment (with preprogrammed numbers) also can improve a plant's emergency communications capability.

For plants handling large quantities of toxic or flammable fluids, field personnel should be able to communicate warnings of leakage to personnel capable of taking mitigating action. (These personnel are usually located in a control room.) In addition to hand-held portable radios, radio transceivers mounted in hard hats allow rapid use and avoid encumbrance. Fixed communications systems involving telephones mounted on building columns or walls may be inaccessible in a vapor-release incident.

Rapid communication of incidents to members of emergency-preparedness teams is essential to their participation in the injury and loss prevention effort. Small radio "beepers" that receive messages from a central location are commonly used for this purpose. Actions to be taken upon sounding of a hazardous vapor alarm are listed in Section 5.3.

7.2 EMERGENCY SHUTDOWN EQUIPMENT AND PROCEDURES

Part of the procedures for every process with vapor cloud release potential must be a well-established emergency shutdown procedure. Although the procedure will be highly specific to a given process, essential elements include

- a clear definition of what constitutes an emergency situation;
- a simple, stepwise statement of the most critical actions to take to bring the emergency under control (i.e., what the "brakes" of the system are, such as shutting off of a critical reactant, dumping the unit, quenching, venting, etc.);
- definition of when an emergency situation should be considered out of control, and appropriate subsequent actions;
- a procedure for obtaining supervisory or technical assistance, with lines of authority and alternate persons to be contacted when a given person is unavailable; and
- on-site communications (as discussed in the preceding section) and off-site communications (as discussed in chapter 8).

In addition to emergency shutdown procedures, various process and auxiliary equipment can be considered to be dedicated to emergency shutdown purposes. This would include not only disposal, isolation, and transfer systems (Chapter 3), but also other equipment necessary to mitigate the release effects, such as automatic closing of control room air intakes and shut-off of the ventilation system (Section 7.4).

7.3 SITE EVACUATION

If a release of flammable vapor occurs, evacuation is the proper response for personnel near the release point and downwind of the release, as long as the evacuation route does not take personnel through the vapor cloud (in which case sheltering may be safer). Only personnel in buildings designed to resist the effects of external fire or explosion should plan to remain, and then only briefly to initiate orderly shutdowns. The guidelines concerning evacuation from toxic vapor releases also apply to personnel evacuating in response to flammable vapor releases.

For releases of toxic or flammable-toxic vapors, personnel downwind from the release point and out-of-doors should leave the area by walking crosswind toward havens or areas not likely to be affected by the release. This may be easily accomplished for visible vapor clouds; however, crosswind evacuation from invisible clouds or from releases at night with poor lighting (and particularly from vapors without warning odors) requires knowledge of both wind direction and direction to the source of release. To provide an indication of wind direction, wind socks, weather vanes, or other devices should be installed in sufficient numbers so that at least one can be seen from any point on the site. These devices also should be illuminated at night, especially in congested areas.

Personnel must be sufficiently familiar with the orientation of the site and the location of potential release sources to recognize the meaning of coded alarms or telephone warnings. Wind socks should be equipped with direction indicators (north/south as a minimum), and maps with alarm codes must be strategically posted (for example, near doorways) throughout the site.

Personnel who are indoors when a plant fume alarm sounds should first isolate the building from the outside environment by shutting off air intakes, and then closing doors and windows. If the vapor cloud has not yet reached the building, the occupants should leave the building and walk crosswind to a safe area.

In congested areas and multistory buildings, the preferred evacuation routes should be clearly marked, particularly where such routes may pass over roofs or other structures, along catwalks, through doorways or halls, and up or down stairs to ground level.

7.4 HAVENS

Havens provide protection against vapor clouds by protecting against thermal radiation for flammable vapor clouds and by having a reduced vapor concentration indoors for toxic vapor clouds. Particularly for flammable vapor clouds, seeking a haven should have priority over trying to escape through the cloud.

Most buildings (and even vehicles) can serve as havens for protection against toxic vapor clouds (Purdy and Davies, 1985), provided that the duration of the cloud is not too great and that interchange of air with the outside can be prevented. Essentially, any weather-tight building makes a reasonable temporary haven. Short, low-concentration puffs usually do not result in high concentrations of vapor in such enclosures. However, prolonged envelopment and the resulting infiltration may require rescue. People within these havens should be notified to go outside as soon as possible after the vapor cloud has passed to avoid prolonged exposure to low-level contaminated air trapped inside.

For greater protection against releases of toxic gases or vapors, occupied structures (particularly interior rooms) can be converted into more effective temporary or permanent havens, depending on the possible duration of releases. Personnel in temporary havens likely would be rescued by well-equipped teams in the event of any prolonged release.

For either temporary or permanent havens, the influx of contaminated air from outside should be minimized by closing intakes for air-handling systems and disengaging exhaust-ventilation systems. For buildings close to sources of potential leakage, these actions should be automatic and actuated by gas sensors in the inlet ducting. In all other buildings on sites handling toxic vapors, switches or push buttons to accomplish these actions should be made readily accessible. Buildings very close to such sources generally should be air-conditioned and have windows and other openings (eaves, cable trays, louvers, etc.) sealed closed; self-closing doors should be installed.

Permanent havens should have all of the above features and should be equipped with air locks. Air intakes should be from widely separated sources or through filters of adequate contaminant-removal capacity. Control rooms and locations from which rescue would be difficult or delayed should be equipped with breathing apparatus (breathing-air cylinders with airline masks, or self-contained breathing apparatus) for the maximum number of personnel likely to be in that haven (Chemical Industries Association, 1979). Appendix E provides guidelines concerning the maximum occupancy for rooms and buildings. The occupancy estimates are based primarily on the increase in humidity developed by persons in confined spaces.

7.5 ESCAPE FROM VAPOR CLOUD

Escape is an action taken when an exposed individual is caught within a portion of the cloud that contains a significant concentration of contaminant. For a flammable vapor release, the only viable response is very rapid crosswind travel to an area unlikely to be affected by the release (Purdy and Davies, 1985). On some sites, the only exception to plant rules prohibiting running is escape from flammable vapors (or fire).

For toxic vapor releases, an alternative to crosswind travel to an unaffected area (Lees, 1980, p. 1035) is entry into a haven (Section 7.4). Selection of the proper escape alternative requires knowledge of the odor-warning properties of the contaminant, the size of the release, wind direction (from wind socks or visible steam, smoke, or vapor plumes), direction to the source, and the probability of rescue from a temporary haven. For most situations, entry into any structure which appears secure against infiltration would be the proper choice. In the absence of a haven, crosswind travel to uncontaminated air should be attempted, but at night or in congested areas escape may be difficult (Purdy and Davies, 1985).

A different situation exists for personnel inside a temporary haven that is being infiltrated with contaminant from outside. Many sites handling toxic gases or vapors provide escape masks (with 5 minutes of compressed breathing air) in operating buildings, in sufficient quantity for all personnel likely to be there (including the occupants of meetings rooms). Note, however, that air-supplied respirators may not provide sufficient protection against those toxic vapors that are absorbed through the skin (e.g., hydrogen cyanide). A much less desirable alternative is providing cartridge respirators for escaping personnel. The undesirable features of respirators are that (1) the concentration outside may be unknown, and the use of cartridge masks is usually limited to known concentrations; and (2) the periodic negative pressure in cartridge masks may result in significant infiltration (and possible panic). Even less desirable alternatives--but possible last resorts-- include covering the head and body with a transparent bag or using wet towels (if the contaminant is soluble in water).

7.6 PERSONAL PROTECTIVE EQUIPMENT

Personal protective equipment provides barriers between a hazardous event and the persons on-site who may be exposed to its effects. Although these barriers may be overwhelmed in severe cases or disabled by improper use, they can be an effective mitigation measure in most emergency situations.

The selection of personal protective equipment to be used on a particular site (or areas of a site) depends on the nature of the potential hazards involved. For fire hazards, fire-retardant clothing has been demonstrated to effectively reduce the hazards of exposure to thermal radiation (fire). For explosion hazards, no specific type of protective clothing can be specified, although hard hats may provide some protection against structural damage and falling debris and hearing protection will minimize the likelihood of eardrum damage resulting from blast overpressures.

For toxic release hazards, respiratory protection is often provided for escape purposes, with many varieties of respiratory protection being available depending on the chemical and the degree of protection required. Mouthpiece or face-mask respirators using cartridges are usually chemical specific and only effective at low to moderate concentrations; however, they can be easily worn on the belt or around the neck, for easy use in an emergency, and are relatively inexpensive. Cartridges should be replaced periodically and whenever the unit is exposed to vapors. Persons equipped with cartridge respirators must clearly understand the limitations of such units. Mask-type or hood-type escape respirators operating under positive pressure from a small "5-minute" compressed-air cylinder are generally more effective in emergency situations because they are not chemical specific and are effective at higher vapor concentrations. Although they also can be worn on the belt, they are heavier to carry and require more training than cartridge-type units. Full face masks with breathing air from an uncontaminated source can be used in a stationary location (e.g., for maintenance work when a pipe flange is opened, for making and breaking tank car connections, or in a control room during an emergency); however, they cannot be used by themselves for escape purposes.

A self-contained breathing apparatus (SCBA) can be used during emergency response actions (e.g., closing a manual isolation valve in an area where hazardous vapors are present), and full protective clothing such as an acid suit may often be required to be used in such a situation. SCBA units are often kept in control rooms and various strategic locations around a plant handling volatile toxic chemicals to allow, for example, control room personnel additional time to safely shut down a process and evacuate the area. It should be noted, however, that in incidents where a dense plume suddenly has been generated and infiltrated an occupied building, the personnel inside the building have not always had time to don breathing protection such as

SCBA, because of the time required to put on the gear, as well as possibly reduced visibility and a tendency to panic.

Human factors must be considered when specifying protective equipment if they are expected to be used without exception. Some considerations are heat stress and humidity (e.g., ability of fire-retardant clothing to "breathe"), weight, appearance, claustrophobia, necessity to shave facial hair, and general ease of use. The site's ongoing training program and safety attitude are the key elements in ensuring effective and continued use of personal protective equipment.

7.7 MEDICAL TREATMENT

For many toxic vapors, particularly those causing respiratory paralysis, prompt medical attention or first aid can reduce the severity of exposure. Rescue teams should be trained in cardiopulmonary resuscitation and be provided with breathing and resuscitation equipment (automatic oxygen resuscitators) and other appropriate rescue equipment (NSC, 1969).

The appropriate treatment for exposure to all hazardous materials at a site should be determined in advance of any emergency. Sufficient treatment facilities should be provided at a medical center or at a location not likely to be involved in a toxic or flammable vapor release. Beds or cots (a minimum of three, and about one per 150 employees), oxygen, showers, and appropriate antidotes should be available.

7.8 ON-SITE EMERGENCY PLANS, PROCEDURES, TRAINING, AND DRILLS

Well-developed emergency plans are an important part of operations involving hazardous materials. The plans should be based on the plant's likely emergencies and tested frequently to assure that the warning and evacuation actions are taken without delay, particularly during periods when fewer personnel are on duty (such as at night and during weekends). Portable radios and computerized dispersion plotters can assist in reducing delays and providing meaningful information to people entrusted with making important (potentially lifesaving) decisions.

For flammable gases and vapors, the primary objectives of emergency plans are to protect potentially exposed personnel (Lees, 1980, pp. 454, 638, 808, and 812) and stop the release. Since most structures on chemical plants are not designed to resist vapor cloud explosions or to protect against fireball combustion outside, taking refuge in buildings until the release has ended (or until rescue) may not be the preferred alternative to evacuation. However, personnel inside buildings at the time of an out-of-doors fire or fireball incident may be partially protected against the initial thermal radiation. In contrast, personnel inside buildings during a vapor cloud explosion (or pressure-vessel explosion) may be subjected to injury from window breakage, ceiling collapse, or structural failure.

For toxic gases and vapors, the primary objectives of emergency plans are to warn the downwind population of impending hazard, so that self-protective measures can be taken, and to stop the release. Most enclosed structures on chemical plants can offer considerable protection against influx of contaminated outside air, and taking refuge in buildings until the release has ended (or until rescue)--with or without additional breathing protection--may be a viable alternative to evacuation.

For flammable toxic gases, liquefied gases, and vapors, the emergency plans should be directed toward evacuation because the likely consequences of release and ignition would be explosions, missiles, or fireballs. For releases of flammable toxic liquids, refuge in havens would be a suitable alternative to evacuation because the consequences of release would be a toxic vapor cloud, possibly accompanied by localized fire at the spill point.

Emergency plans for mitigation of vapor cloud hazards should be developed for all applicable types of releases. Thus, the first step in developing such plans would be to list the materials (reactants, solvents, products, and by-products) found at the site in potentially dangerous quantities. It will also be necessary to identify likely emergencies or specific accident scenarios on which to base the emergency plans (Johnson, 1986). The degree of potential danger will depend primarily on toxicity; for example, a potentially dangerous quantity of propane (essentially nontoxic, but highly flammable) may be of the order of a ton or more (vapor cloud explosion hazard), while release of 100 pounds of methyl isocyanate (IDLH concentration 1 ppm) may create a serious toxicity hazard over a wide area. Emergency plans should include the following features (EPA, 1985; National Response Team, 1987):

- A one-page presentation of the critically important first-step actions to be taken in response to a release of toxic or flammable vapor. This may include the telephone numbers of the plant manager, operating and maintenance superintendents, safety supervisor, local authorities (police and fire department), and nearby hospitals.

- Preplanned messages to authorities, specifying the name of the caller, the company name and telephone number, the material released, the type of hazard (toxic/flammable), the time of release, the wind direction, and the expected duration of release.
- An organization chart showing the persons in charge and the chain of command for all shifts.
- A tabulation of the communications systems available within the site and for communicating with people off site.
- Procedures for personnel accounting, including designated rally spots, communications to be used, and methods for searching for missing employees, contractors, and visitors.
- A listing of the hazardous materials used at the site, together with the physiological effect of exposure (concentration and dose), warning properties, and antidotes or medical treatment procedures. It may also be appropriate to include the quantities stored at the site and the maximum potential leakage rates (for example, pumping rates).
- A listing of the equipment on the site for dealing with hazardous material releases, including detectors, self-contained breathing apparatus (and reserve air cylinders), protective clothing, foam generators, sandbag locations, emergency lighting, flare guns, etc.
- A listing of the buildings at the site, together with an assessment of the protection provided by existing buildings.
- A listing of the medical personnel and facilities available, with hours of operation.
- A map of the site showing roadways into and out of the potential hazard areas so that roadblocks can be set up and access and exit lanes can be kept clear.
- A map of the fire-main system showing hydrants and deluge systems.
- A map showing the location of emergency shut-off valves (automatic and manual) and their identification.
- Plans for dealing with other types of emergencies, such as adjacent-plant emergencies, hurricanes, floods, utility failures, bomb threats.
- Plans for releases which might endanger off-site personnel or property, including notification of authorities and concentrations of population (schools, hospitals, office buildings, apartments, prisons, shopping malls, and other public buildings).

Drills should be scheduled so that every employee participates in at least one drill per year. Thus, it may be necessary to schedule three or four drills per year.

REFERENCES

Chemical Industries Association, 1979: *An Approach to the Categorisation of Process Plant Hazard and Control Building Design*, Chemical Industries Association, London, 25.

EPA, 1985: *Chemical Emergency Preparedness Program--Interim Guidance*, U.S. Environmental Protection Agency, Washington, D.C., 3-6 to 3-11, 4-5, 4-6, 4-9 to 4-16.

Johnson, R. W., 1986: Toxic release scenarios for emergency response plans. Paper presented at 79th Annu. Meet. Air Pollut. Control Assoc., Minneapolis.

Lees, F. P., 1980: *Loss Prevention in the Process Industries*, Butterworth, London.

National Response Team, 1987: *Hazardous Materials Emergency Planning Guide*, NRT-1 (formerly FEMA-10), pp. 39-41, 48, 51-54, and D-9.

NSC, 1969: *Accident Prevention Manual for Industrial Operators*, 6th Ed., National Safety Council, Chicago, 1132.

Purdy, G., and P. C. Davies, 1985: Toxic gas incidents--Some important considerations for emergency planning. *Inst. Chem. Eng. Loss Prev. Bull. 062*, 2.

8

ALERTING LOCAL AUTHORITIES AND THE PUBLIC

Prompt alerting of local authorities has allowed effective action to be taken in many situations where hazardous vapors have been released. The plant's responsibility in an emergency is to quickly notify the proper authorities, and the authorities then have the responsibility to notify the public.

8.1 ALERTING SYSTEMS

8.1.1 Capabilities

Several manufacturers, as listed in Table 8-1 (McNaughton et al., 1986), have combined a computer with a television-like monitor to estimate the dangers to downwind personnel and to communicate emergency plans. Displays show wind direction and nominal dispersion plots based on properties of a typical plant vapor to illustrate the effects of wind direction and atmospheric stability.

Table 8-1. Toxic Spill Alerting Systems

Manufacturer	Address	Telephone	Trade Name
Dow Chemical Company, USA	P.O. Box 150 Plaquemine, LA 70764	504-685-8000	DAISY
E. I. du Pont de Nemours & Company, Inc.	Savannah River Plant Aiken, SC 29808	803-450-6211	WIND
Environmental Research & Technology, Inc.	696 Virginia Road Concord, MA 01742	617-369-8910	HASTE
NUS Corporation	910 Clopper Road Gaithersburg, MD 20878	301-258-6000	EMERGE
Pickard, Lowe, and Garrick, Inc.	1200 18th Street NW Washington, DC 20036	202-296-8633	MIDAS
Radian Corporation	8501 Mo-Pac Highway Austin, TX 78766	512-454-4797	CHARM
Safer Emergency Systems, Inc.	5700 Corsa Avenue Westlake Village, CA 91362	818-707-2777	SAFER

These systems are particularly useful in training programs and in accident scenario development. They vary considerably in capability and need to be thoroughly understood before attempting "real-time" use in assessing the consequences of a release or in projecting future personnel exposures.

Since computation time for some systems approaches one minute, it is desirable to continuously compute and display a nominal case for all systems for rapid initial response in an emergency. As information is received concerning release location, size, and material, the inputs are revised to present a realistic portrayal of the probable sequence of events.

8.1.2 Input Requirements

Wind direction information should be constantly fed into the system. For real-time dispersion calculations, wind-speed input is also required. Temperature and time inputs are used to estimate atmospheric-stability class and rates of vaporization from pools. Relative humidity may also be used to estimate the potential hazard of some aerosol-forming acid vapors.

Additional information on input data required for modeling, as well as evaluation of the various codes used for computer calculations of plume dispersion, is given in the AIChE-CCPS document *Guidelines for Use of Vapor Cloud Dispersion Models* (1987).

8.2 ROLES AND LINES OF COMMUNICATION

The objective of emergency planning for off-site exposure is to communicate credible hazard warnings (Ikeda, 1982) and appropriate mitigation actions to people outside the site boundary in an emergency. Clear roles and lines of communication therefore should be established with public authorities, hospitals, the public, and the media. Of course, prevention of vapor releases is an even more important aspect of process control and environmental protection (AIChE-CCPS, 1985, preface).

The likelihood of the public and authorities accepting recommendations concerning actions to take in advance of danger (credibility of warnings) is a major concern of the Chemical Manufacturers Association's CAER (Community Awareness and Emergency Response) program (CMA, 1985, pp. 12, 27). The public's familiarity with a chemical plant's processes, potential for accidental releases, and emergency plans for off-site populations can have a favorable effect on a community's acceptance of forewarning announcements (Dow, 1986).

The CAER program encourages drills involving plant personnel and public authorities. These drills should use the communications systems and messages specified in the emergency plan. Only through such drills can deficiencies in planning be identified so that prompt and effective action can be taken in any vapor-release emergency.

8.3 INFORMATION TO BE COMMUNICATED

The first step in planning for protection of off-site populations is to determine the minimum release size and release rate that would be dangerous beyond the site boundary. These values should be determined for all of the hazardous materials handled at the site.

Detection of serious leakage of any material identified as being within the scope of the emergency plans must be followed quickly by an estimation of leak size. Potentially exposed areas of the population are then identified. Computerized alerting systems (Section 8.1) are available (McNaughton et al., 1986; Basta, 1985) to streamline the steps between estimation of leak size and communication with the exposed population and public authorities. Some computer systems that track vapor releases on a monitor also display lists of telephone numbers to aid in rapidly issuing warnings to areas at risk.

The following information should be accurately communicated in an emergency:

- materials released,
- estimated amount of release,
- direction of release,
- current meteorological conditions, and
- estimated concentrations at particular distances and locations.

Other aspects of emergency response planning and hazard communications are described in the *CAER Program Handbook* (CMA, 1985).

REFERENCES

Basta, D. H., 1985: U.S. CPI to feel effects of Bhopal tragedy. *Chem. Eng. 92* (6), 27.

AIChE-CCPS, 1985: *Guidelines for Hazard Evaluation Procedures*, prepared by Battelle Columbus Division for American Institute of Chemical Engineers--Center for Chemical Process Safety, New York.

AIChE-CCPS, 1987: *Guidelines for Use of Vapor Cloud Dispersion Models*, prepared by S. R. Hanna and P. J. Drivas for American Institute of Chemical Engineers--Center for Chemical Process Safety, New York.

CMA, 1985: *Community Awareness and Emergency Response Program Handbook*, Chemical Manufacturers Association, Washington, D.C.

Dow, 1986: *Community Awareness/Emergency Response*, Dow Chemical Company, Freeport, TX, videotape 86-1017.

Ikeda, K., 1982: Warning of disaster and evacuation behavior in a Japanese chemical fire. *J. Hazard. Mater. 7* (1), 51.

McNaughton, D. A., et al., 1986: *Evaluation and Assessment of Models for Emergency Response Planning*, prepared by TRC Environmental Consultants, Inc. for Chemical Manufacturers Association, Washington, D.C.

9

SELECTION OF MITIGATION MEASURES

TO: Plant Manager
RE: Vapor Release Mitigation

Which mitigation measures will you be using at your site?

A. Reduce storage inventory   •
B. Substitute hydrate for nitrate   •
C. Replace distillation column with higee   •
D. Buy up land around site   X. All of the above
E. Change materials of construction Y. None of the above
F. Scrub all relief valve releases   Z. Risk analysis in progress
G. Install remote isolation valves
H. Build dike around storage tank
I. Beef up training program

Three items of information needed for the selection of mitigation measures are as follows:

- the risk (type and magnitude) that exists without mitigation,
- the risk reduction that can be achieved by each of the candidate mitigation measures, and
- the cost and feasibility of each candidate measure.

Knowing the existing risk of each potential accidental release, one is able to allocate resources where they will have the greatest beneficial impact. Knowing how much risk reduction is possible and the costs of each alternative, one is able to choose the most appropriate mitigation approach for a particular situation.

Risk is a function of the probability that an accidental event will occur and of the undesirable consequences if it does occur. However, frequently it is considered more important to reduce the consequences of a potential accident than to reduce its probability even though the reduction in risk might be the same. The forthcoming AIChE-CCPS document *Guidelines for Chemical Process Quantitative Risk Assessment Procedures* will discuss this in greater detail.

To determine the existing risk that a design or operating facility presents it is necessary to identify and evaluate all of the existing hazards and all of the potential accidental events (accident scenarios) that make up that risk (AIChE-CCPS, 1985). The risk reduction that is possible by a particular mitigation measure can then be determined by repeating the evaluation with the candidate mitigation measure in place.

9.1 RISK ANALYSIS

If the design or facility is a replica of many identical systems and there has been sufficient operating experience and accumulation of data on accidents and plant upsets for those systems, many of the hazards and accidental events, as well as the risk that each represents, will be known. If the effectiveness of the mitigation measures under consideration is also known, the selection will be relatively easy. More often, the systems are not identical, and sufficient data and experience on the facility and mitigation measures are not available. If that is the case, it will be necessary to examine the facility using recognized methods of risk analysis

- to identify the hazards and potential accidents;
- to estimate the undesirable consequences of the potential accidents; and
- to predict the probability of equipment failures, human errors, and external events that could cause the accidents.

The risk analysis may take the form of the flow chart (AIChE-CCPS, 1985) shown in Figure 9-1. This series of steps starts with hazard identification and proceeds from very qualitative estimates of consequences, probability, and the effectiveness of candidate mitigation measures to highly quantitative evaluation of probability and consequences, if that is necessary. It permits terminating the analysis at any point that qualitative estimates are sufficient to determine if mitiga-

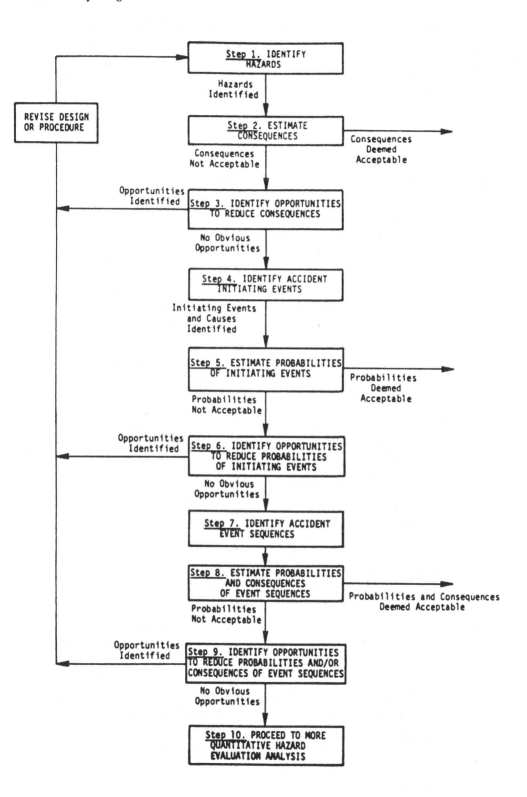

Figure 9.1. *Risk Analysis Sequence*

tion is needed and which measures to select. If qualitative analysis does not provide the degree of certainty needed, the analysis can continue with quantitative analysis, building on the knowledgegained from the previous steps. At each step there are risk analysis methods that have been accepted by the chemical process industry that can be used to identify and evaluate hazards, accidents scenarios, probabilities, and consequences.

9.2 METHODS FOR HAZARD IDENTIFICATION

The AIChE-CCPS document *Guidelines for Hazard Evaluation Procedures* (1985) describes four methods for identifying hazards and for qualitatively identifying accident causes and accident event sequences:

- *Preliminary Hazards Analysis*--useful for recognizing hazards at an early stage of plant or system design.
- *"What If" Analysis*--examines possible deviations from the design/operating intent through asking "What If" questions. This method is somewhat less structured than other procedures. It is most useful if the team has experience with the system being analyzed.
- *Hazard and Operability (HAZOP) Study*--identifies process deviations that could lead to undesirable consequences by applying a specific set of "guide words" to process parameters. The HAZOP study employs a multidisciplinary team approach.
- *Failure Modes, Effects, and Criticality Analysis*--examines specific failure modes of process and control equipment and semiquantitatively evaluates the effects of such failures.

9.3 METHODS FOR ESTIMATING THE CONSEQUENCES OF ACCIDENTS

Whereas logic models (fault trees and event trees) can be used to determine the probability of accidents involving complex event sequences, consequence models are used to model the physical and chemical behavior of complex engineered systems under upset conditions and determine the consequence severities of resulting accidents. The physical models may be detailed and complex or may be simple approximations of what is likely to happen, depending upon the degree to which precise estimates are needed. Often a "worst case" approach is used for a first approximation to get an upper limit to the consequences of a particular type of accident. If the consequences so determined can be accepted by all parties affected, a worst case analysis is sufficient. However, if the consequences are too uncertain or too severe to be acceptable, a more refined analysis may be in order.

Physical models have been developed to describe the initiation and progression of fires, the detonation of explosive materials and production and effects of projectiles and shrapnel, and the release and dispersion of toxic and flammable gases. Evaluating the hazards of flammable-gas releases also involves the possible ignition and explosion of the gases at some point along the path of the vapor cloud. This document concentrates on vapor cloud releases; thus, dispersion models are the consequence analysis methods of most interest. However, since fires and/or explosions are possible consequences of vapor cloud releases, fire/explosion analysis may also be of interest when evaluating the effectiveness of mitigation measures.

The undesirable consequences of an accident--above and beyond the loss of plant, equipment, and ability to produce a product--are the effects of the accident on the health of the employees and neighboring public. The physical models used to assess off-site consequences will not directly predict health effects. The physical models will predict concentrations of a toxic or flammable gas and heat radiation levels as a function of distance from the source, height of cloud, time, etc. Data on the effects of exposure to the toxic gas, heat radiation, and/or overpressure are needed to predict health effects.

Some useful information on consequence analysis methods can be found in the following sources:

- An AIChE-CCPS document on vapor cloud dispersion entitled *Guidelines for Use of Vapor Cloud Dispersion Models* (1987). This document includes numerous references to original sources.
- Another AIChE-CCPS document entitled *Guidelines for Chemical Process Quantitative Risk Assessment Procedures*, to be published in 1988. It will describe other methods for quantitative consequence analysis.
- Estimates of in-plant and equipment damage from fires and explosions and employee health effects from exposure to toxic gases, as developed from the Dow and Mond Hazard Indices (AIChE-CCPS, 1985). The indices provide an empirical method of ranking different areas of a plant for hazards and for estimating the costs of plant and equipment losses from accidents. Weighting factors and nomographs are provided, based on experience, which permit quantification without the use of phenomenological models. The Indices are particularly useful in the early stages of plant design, to avoid having particularly hazardous situations develop.

9.4 METHODS FOR ESTIMATING THE PROBABILITY OF ACCIDENTS

The AIChE-CCPS document Guidelines for Hazard Evaluation Procedures (1985) describes three methods for identifying and displaying the relationships between the sequential events of an accident scenario. These methods also provide a means for estimating the frequency or probability of initiating events and accident event sequences. They are as follows:

- *Fault-Tree Analysis*--a deductive, graphic technique for finding basic causes (equipment failures, human errors, external events) of a particular accident event. It is useful for identifying combinations of failures, errors, etc. needed to cause a particular accident. It can be used for calculating the frequency or probability of an accident event sequence.
- *Event-Tree Analysis*--an inductive, graphic technique for describing various outcomes (accident event sequences/accident scenarios) of an initiating event via successes or failures of mitigating measures. It can be used for calculating the probability of each outcome.
- *Cause-Consequence Analysis*--a combination of fault-tree analysis and event-tree analysis.

These methods will be described in greater detail in the forthcoming AIChE-CCPS document entitled *Guidelines for Chemical Process Quantitative Risk Assessment Procedures*.

REFERENCES

AIChE-CCPS, 1985: *Guidelines for Hazard Evaluation Procedures*, prepared by Battelle Columbus Division for American Institute of Chemical Engineers--Center for Chemical Process Safety, New York.
AIChE-CCPS, 1987: *Guidelines for Use of Vapor Cloud Dispersion Models*, prepared by S. R. Hanna and P. J. Drivas for American Institute of Chemical Engineers--Center for Chemical Process Safety, New York.

APPENDIX A

LOSS-OF-CONTAINMENT CAUSES IN THE CHEMICAL INDUSTRY

Plant Inventory Discharged to Environment Due to Loss of Containment

(Note: This cannot presume to be an exhaustive list of causes.)

I. CONTAINMENT LOST VIA AN "OPEN-END" ROUTE TO ATMOSPHERE
   A. Due to genuine process relief or dumping requirements
   B. Due to maloperation of equipment in service, e.g., spurious relief valve operation or rupture disk failure, etc.
   C. Due to operator error, e.g., drain or vent valve left open, misrouting of materials, tank overfilled, unit opened up under pressure, etc.

II. CONTAINMENT FAILURE UNDER DESIGN OPERATING CONDITIONS DUE TO IMPERFECTIONS IN THE EQUIPMENT
   A. Imperfections arising prior to commissioning and not detected before start-up (due to poor inspection or testing procedures)
      1. Equipment inadequately designed for proposed duty, e.g., wrong materials specified, pressure ratings of vessel or pipework inadequate, temperature ratings inadequate, etc.
      2. Defects arising during manufacture, e.g., wrong materials used, poor workmanship, poor quality control, etc.

3. Equipment damage or deterioration in transit or during storage
4. Defects arising during construction, e.g., welding defects, misalignment, wrong gaskets fitted, etc.

B. Imperfections due to equipment deterioration in service and not detected before the effect becomes significant (due to inadequate monitoring procedures in those cases where deterioration is gradual)
   1. Normal wear and tear on pump or agitator seals, valve packing, flange gaskets, etc.
   2. Internal and/or external corrosion, including stress corrosion cracking
   3. Erosion or thinning
   4. Metal fatigue or vibration effects
   5. Previous periods of gross maloperation, e.g., furnace operation at above the design tube skin temperature ("creep")
   6. Hydrogen embrittlement

C. Imperfections arising from routine maintenance or minor modifications not carried out correctly, e.g., poor workmanship, wrong materials, etc.

III. CONTAINMENT FAILURE UNDER DESIGN OPERATING CONDITIONS DUE TO EXTERNAL AGENCIES
   A. Impact damage, such as by cranes, road vehicles, excavators, machinery associated with the process, etc.
   B. Damage by confined explosions due to accumulation and ignition of flammable mixtures arising from small process leaks, e.g., flammable gas build-up in analyzer houses, in enclosed drains, around submerged tanks, etc.
   C. Settlement of structural supports due to geological or climatic factors or failure of structural supports due to corrosion, etc.
   D. Damage to tank trucks, rail cars, containers, etc., during transport of materials on- or off-site
   E. Fire exposure
   F. Blast effects from a nearby explosion (unconfined vapor cloud explosion, bursting vessel, etc.), such as blast overpressure, projectiles, structural damage, etc.
   G. Natural events (acts of God) such as windstorms, earthquakes, floods, lightning, etc.

IV. CONTAINMENT FAILURE DUE TO DEVIATIONS IN PLANT CONDITIONS BEYOND THE DESIGN LIMITS

A. Overpressuring of equipment
   1. Due to a connected pressure source
      a. gas pressure source
         (1) gas breakthrough into downstream low-pressure equipment due to failure of a pressure or level controller, isolation valve opened in error, etc.
         (2) pressurized backflow into low-pressure equipment, e.g., due to compressor failure
      b. liquid pressure source
         (1) pumping up of blocked-in gas spaces
         (2) hydraulic overpressuring due to a block-in condition downstream
         (3) excessive surge or hammer, such as by sudden valve closure on liquid transfer line
   2. Due to rising process temperature
      a. loss of cooling
         (1) loss of coolant flow, e.g., to a reactor cooler, to a distillation column condenser, etc.
         (2) elevated coolant temperature, e.g., loss of cooling water fans, etc.
         (3) fouling of coolers, condensers, or exchangers
      b. excessive heat input (thermal)
         (1) heater control faults, such as on steam or hot oil heated systems
         (2) ingress of hot extraneous materials, e.g., slop-over
      c. excessive heat generation (chemical)
         (1) reactor runaway, e.g., due to loss of reaction diluent, high feed rate, high molar ratio, accumulation of unreacted reactants due to inadequate mixing or temporary loss of reaction subsequently leading to a runaway, etc.
         (2) exotherming due to ingress of catalytic impurities, e.g., backflow from ethylene oxide consumer unit into feed tank
         (3) exotherming due to mixing of incompatible chemicals, e.g., H_2SO_4 with NaOH
         (4) exothermic decomposition of thermally unstable or explosive material such as peroxides, e.g., due to temperature rise, overconcentration, or deposition on hot surfaces

3. Due to an internal explosion arising from formation and ignition of flammable gas mixtures, mists, or dusts
   a. ingress of air, e.g., due to inadequate purging of equipment at plant start-up, due to loss of nitrogen purge on flare headers, storage tanks, centrifuge systems, dryers, etc.
   b. loss of critical inert diluent, e.g., loss of nitrogen padding on an ethylene oxide storage tank, loss of nitrogen to the make-up section of a nitrogen/air solids conveying system
   c. failure of explosion suppressants
   d. flammable excursion in oxidation processes, e.g., due to high air or oxygen rates, or loss of conversion
4. Due to physically or mechanically induced forces or stresses
   a. expansion upon change of state, e.g., freezing of water in pipe runs
   b. thermal expansion of blocked-in liquids, e.g., in heat exchangers or long pipe runs
   c. ingress of extraneous phases, e.g., gas compressor failure due to liquid carry-through to machine suction, condensate hammer in steam lines, etc.

B. Underpressuring of equipment (for equipment not capable of withstanding vacuum)
   1. By direct connection to an ejector set or to equipment normally running under vacuum
      a. due to equipment malfunction, e.g., loss of liquid seal due to failure of a level controller causing vacuum to be applied upstream, etc.
      b. due to operator error, e.g., isolation valve left open, etc.
   2. Due to the movement or transfer of liquids
      a. pumping out of tanks or vessels
      b. emptying or draining elevated blocked-in equipment under gravity
   3. Due to cooling of gases or vapors
      a. condensation of condensable vapors, e.g., vessel blocked-in after steaming
      b. cooling of noncondensable gases or vapors, e.g., storage tank by heavy from rainfall in summer
   4. Due to solubility effects, e.g., dissolution of gases in liquids

C. High metal temperature (causing loss of strength)
   1. Fire under equipment, e.g., due to spillage, pump leak, etc.
   2. Flame impingement causing local overheating, e.g., on furnaces due to misalignment or maladjustment of burners
   3. Overheating by electric heaters, e.g., due to failure of high temperature cutout
   4. Inadequate flow of fluid via heated equipment, e.g., furnace tube failure on loss of hot oil flow
   5. Higher flow rate or higher temperature of the hotter stream, or lower flow rate or higher temperature of the colder stream, via a heat exchanger
D. Low metal temperature (causing cold embrittlement and overstressing)
   1. Overcooling by refrigeration units, e.g., due to control faults, wrong refrigerant, etc.
   2. Incomplete vaporization and/or inadequate heating of refrigerated material before transfer into equipment of inadequate temperature rating, e.g., due to control faults on a liquid ethylene vaporization unit
   3. Loss of system pressure on units handling liquids of low boiling point
E. Wrong process materials or abnormal impurities (causing accelerated corrosion, chemical attack of seals or gaskets, stress corrosion cracking, embrittlement, etc.)
   1. Variations in stream compositions outside design limits
   2. Abnormal impurities introduced with raw materials or wrong raw materials
   3. By-products of abnormal chemical reactions
   4. Oxygen, chlorides, or other impurities remaining in equipment at start-up due to inadequate evacuation or decontamination
   5. Impurities entering process from atmosphere, service connections, tube leaks, etc., during operation

APPENDIX B

PROPERTIES OF SOME HAZARDOUS MATERIALS[a]

Material	Molecular Weight	LFL (%)	Flash point (°C)	Water Soluble?	SPEGL (ppm)	TLV[b] (ppm)	IDLH (ppm)
Acetic anhydride	102	2.7	49	Yes	--	5	1,000
Acetone	58	2.5	-20	Yes	--	750	20,000
Acetonitrile	41	3.0	6	Yes	--	40	4,000
Acrolein	56	2.8	-26	Yes	--	0.1	0.5
Acrylonitrile	53	3.0	0	Yes	--	2	4,000
Ammonia	17	15.	Gas	Yes	--	25	500
Aniline	93	1.3	70	No	--	2	100
Arsine	78	Decomp.	Gas	No	--	0.05	6
Benzene	78	1.3	-11	No	--	10	2,000
Bromine	160	NF	NF	No	--	0.1	10
Butadiene (1,3)	54	2.0	Gas	No	--	10	20,000
Butane	58	1.6	Gas	No	--	800	--
Carbon monoxide	28	12.5	Gas	No	--	50	1,500
Carbon disulfide	76	1.3	-30	No	--	1	25
Chlorine	71	NF	NF	No	--	1	25
Chloroform	119	NF	NF	No	--	10	1,000
Cyclohexane	84	1.3	-20	No	--	300	10,000
Dimethylamine	45	2.8	Gas	Yes	--	10	2,000
Dimethyl sulfate	126	(4)	83	No	--	0.1	10
Ethane	30	3.0	Gas	No	--	--	--
Ethyl alcohol	46	3.3	13	Yes	--	1,000	--
Ethylene	28	2.7	Gas	No	--	--	--
Ethylene oxide	44	3.0	-29	Yes	--	1	800
Formaldehyde	30	7.0	Gas	Yes	--	1	100
Gasoline	114	1.4	-43	No	--	300	--
Hydrazine	32	2.9	38	Yes	0.12[c]	0.1	80

Material	Molecular Weight	LFL (%)	Flash point (°C)	Water Soluble?	SPEGL (ppm)	TLV[b] (ppm)	IDLH (ppm)
Hydrogen	2	4.0	Gas	No	--	--	--
Hydrogen chloride	37	NF	NF	Yes	--	5	100
Hydrogen cyanide	27	5.6	-18	Yes	--	10	54
Hydrogen fluoride	20	NF	NF	Yes	--	3	20
Hydrogen sulfide	34	4.0	Gas	Yes	--	10	300
Methane	16	5.0	Gas	No	--	--	--
Methyl alcohol	32	6.0	11	Yes	--	200	25,000
Methyl amine	31	4.9	Gas	Yes	--	10	100
Methyl chloride	51	8.1	-46	No	--	50	10,000
Methyl mercaptan	48	3.9	Gas	Yes	--	0.5	400
Nitrogen dioxide	46	NF	NF	Yes	1	3	50
Phenol	94	1.8	79	Yes	--	5	100
Phosgene	99	NF	NF	Yes	--	0.1	2
Phosphine	34	--Pyrophoric--		No	--	0.3	200
Phos. trichloride	137	NF	NF	Yes	--	0.2	50
Propane	44	2.1	Gas	No	--	--	20,000
Sulfur dioxide	64	NF	NF	Yes	--	2	100
Tetrahydrofuran	72	2.0	-14	Yes	--	200	20,000
Toluene	92	1.2	4	No	--	100	2,000
Vinyl acetate	86	2.6	-6	No	--	10	--
Vinyl chloride	63	3.6	Gas	No	--	5	--

[a] Abbreviations: LFL, lower flammable limit; SPEGL, short-term public emergency guidance level (60-minute); TLV, threshold limit value; IDLH, immediately dangerous to life and health (30-minute); NF, not flammable.

[b] Note: ethane, ethylene, hydrogen, methane, and propane act primarily as simple asphyxiants (besides possible fire/explosion hazards) when present in high concentrations in air; that is, the limiting factor is the available oxygen. The minimal oxygen content should be 18 percent by volume under normal atmospheric pressure (ACGIH, 1986).

[c] SPEGL of 0.12 ppm is for hydrazine; SPEGL=0.24 ppm for methyl hydrazine and 1,1-dimethylhydrazine.

REFERENCES

ACGIH, 1986: *Threshold Limit Values and Biological Exposure Indices*, American Conference of Governmental Industrial Hygienists, Cincinnati, OH, 6, 48.

Merck, 1983: *The Merck Index*, 10th Ed., Merck and Co., Rahway, NJ.

National Academy of Sciences, 1984-1987: *Guidance Levels for Emergency and Continuous Exposure to Selected Airborne Contaminants*, 7 Vols., National Academy of Sciences, National Research Council, National Academy Press, Washington, D.C.

NFPA, 1984: Fire hazard properties of flammable liquids, gases, and volatile solids. *National Fire Code*, NFC 325M, National Fire Protection Association, Boston.

NIOSH, 1985: *Pocket Guide to Chemical Hazards*, DHHS Publication No. 85-114, U.S. Department of Health and Human Services-National Institute for Occupational Safety and Health, Washington, D.C.

APPENDIX C

DERIVATION OF FOG CORRELATIONS:
Correlations between Vapor Concentration, Relative Humidity, and Heat of Solution

Tests have been conducted recently (Schotte, 1987a, 1987b) to determine the aerosol-generating properties of anhydrous hydrogen chloride and hydrogen fluoride. The results of these tests and other observations (Stephenson, 1970) can be interpreted to yield plots of concentration versus relative humidity to divide regions of aerosol fog generation from regions where fog is not generated.

The data can be used further to develop a correlation of concentration C (ppm) and relative humidity RH (%) based on heat of solution H_s (calories per gram-mole), in an equation of the form

$$C = 1 \times 10^6 \left[1 - (RH/100)^k \right]$$

where

$$k = \frac{0.0166}{H_s^{2/3}}$$

or in an equation of the form

$$C = K \ln (100/RH)$$

where

$$K = \frac{1.75 \times 10^6}{H_s^{2/3}}$$

Figure C-1 shows the behavior of hydrogen chloride and hydrogen fluoride, together with the expected behavior of sulfur trioxide, in generating fog from humid air. These curves could be used to estimate the concentration cf vapor at the edge of a cloud (if the relative humidity is known) and then to estimate the size of the spill (from the maximum downwind distance to the visible edge of the cloud and its width), using dispersion equations.

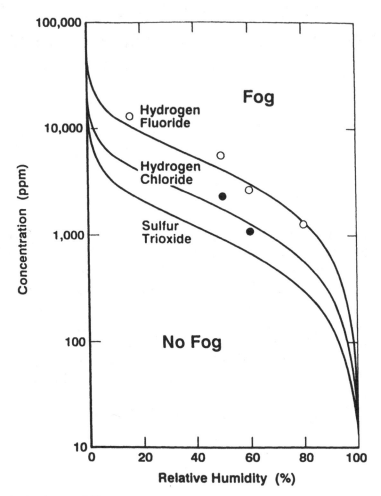

Figure C-1. *Aerosol Fogs*

REFERENCES

Schotte, W., 1987a: Adiabatic expansion of hydrogen chloride and fog in air. *Ind. Eng. Chem. Res. 26* (1), 134.

Schotte, W., 1987b: Fog formation of hydrogen fluoride in air. *Ind. Eng. Chem. Res. 26* (2), 300.

Stephenson, F. G. (ed.), 1970: *Properties and Essential Information for Safe and Use of Hydrofluoric Acid*, Manufacturing Chemists Association Safety Data Sheet SD25, Chemical Manufacturers Association, D.C.

APPENDIX D

CATCHTANK DESIGN

In a recent article by Grossel (1986), design and sizing details are given for emergency relief system catchtanks. This appendix adds supplementary (new) material to the catchtank design information in the article (Grossel, 1987) on two topics.

Vortex Separator. Recently, Professor Mayinger of the Technical University of Munich presented preliminary results of his investigation of the design criteria and performance characteristics of a new type of vortex separator which can be used, like a catchtank, as a vapor-liquid separator (Muschelknautz and Mayinger, 1986). As can be seen from Figure D-1, one obvious advantage of this type of separator is that it can be mounted directly above a reactor so that the vent line between the reactor and separator is vertical with no elbows. The guide blades (vanes) create a vortex motion so that the liquid is deposited on the walls of the conical section and flows out through the annular area beween the vapor line and the conical section. An angle of attack of 30° of the guide blades seemed to provide the best compromise between separation efficiency and low pressure drop. Because the liquid is separated from the vapor-liquid mixture vented during a runaway, the relief device can be smaller in size than would be required if it was mounted on the reactor.

Sizing and Design of a Quencher Catchtank/Knock-out Drum: Quencher Arm and Pool Section. The necessary quantity of quench fluid can be calculated by the following approximate formula (Fauske, 1986):

$$m = \frac{m_0 \, (T_R - T_a) \, C_R}{(T_a - T_0) \, C_q}$$

where m_0 is the mass of reactants, T_R is the temperature of reactants at relief set pressure, T_a is the allowable temperature following complete quench, T_0 is the initial temperature of the quench fluid, and C_q and C_R are the specific heat of the quench fluid and reactants, respectively (in consistent English or SI units).

REFERENCES

Fauske, H., 1986: Disposal of two-phase emergency releases. Paper presented the 4th Miami International Symposium on Multi-Phase Transport and Phenomena, Miami Beach, December 15-17.

Grossel, S. S., 1986: Design and sizing of knock-out drums/catchtanks for emergency relief. *Plant/Oper. Prog.* 5 (3), 129.

Grossel, S. S., 1987: Personal communication, August. Muschelknautz, S., and F. Mayinger, 1986: Fluid separation in two-phase flow depressurization. Paper presented at the 1986 European Two-Phase Meeting, Munich, June 10-13.

Figure D-1. *Vortex Separator*

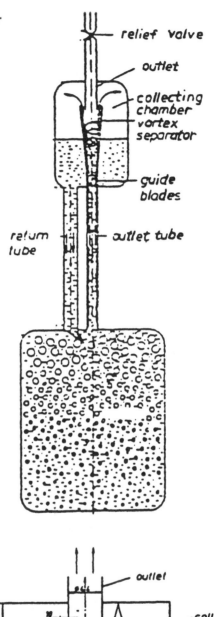

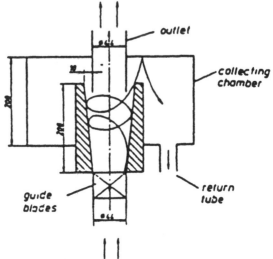

APPENDIX E

CAPACITY OF HAVENS

A haven is essentially a place of refuge, sometimes designed as a sealed enclosure with no interchange of air with the surrounding environment (outside air, or air in adjacent rooms, corridors, or spaces above or below the haven). The interchange is prevented by sealed openings, tight-fitting doors (and/or air locks), and tight-fitting sealed windows, with no intake or exhaust ventilation (Chemical Industries Association, 1979). As discussed below, air conditioning of a haven is desirable (to remove heat and humidity), provided that no interchange of air occurs at the air-conditioning equipment. The purpose of this appendix is to establish the number of cubic feet of space required per person in a haven (or the number of square feet of floor area with a given ceiling height).

The conditions which govern the required volume per person are:

- the oxygen concentration in the haven should not decrease below 18 percent by volume (ACGIH, 1986);
- the carbon dioxide concentration in the haven should not exceed 3 percent by volume (30,000 ppm) for a 15-minute haven occupancy (ACGIH, 1986);
- the temperature should not exceed 92°F (33°C) for persons standing while doing no work, at 100 percent relative humidity (ACGIH, 1986); and
- the volume of the human body is about 75 liters (for a 165-pound person) or about 2.65 cubic feet.

The approximate physiological characteristics of humans which would necessitate limiting the number of persons in a haven (requiring a minimum number of cubic feet per person) are:

- *Respiration (exchange of carbon dioxide for oxygen) (Guyton, 1966).*

Gas or Vapor	Inhaled Air	Exhaled Air
Nitrogen	78.62 v%	74.5 v%
Oxygen	20.84 v%	15.7 v%
Carbon Dioxide	0.04 v%	3.6 v%
Water	0.50 v%	6.2 v%
TOTALS	100.00 v%	100.0 v%

From a nitrogen balance, the volume of exhaled air is about 5.5 percent greater than the inhaled air. Alveolar ventilation is about 350 milliliters per breath and a breathing rate (seated, with moderate stress) would be about 15 breaths per minute (Guyton, 1966). Thus,

Gas or Vapor	Volume (liters per minute)
Nitrogen	0.00
Oxygen	0.21 consumed
Carbon Dioxide	0.21 generated
Water	0.32 generated

Since the energy per liter of oxygen consumed (or carbon dioxide generated) is about 4.825 kcal (Guyton, 1966), the above rates correspond to a metabolic rate of about 1.0 kcal per minute.

- *Generation of Heat.* For persons seated at rest (on chairs or on the floor) at a temperature of 92°F (33°C), about 0.4 kcal per minute of sensible heat would be generated per person, and about 1.0 kcal per minute of latent heat (moisture from breathing and perspiration) would be generated (Baumeister et al., 1987). About 50 percent of body heat would be radiated to the ceiling, walls, and floor of the haven (Guyton, 1966); that is, about 0.2 kcal per minute would be absorbed by the air in the haven, raising the air temperature.

- *Generation of Moisture.* The minimum rate of moisture generation corresponds to a heat loss of about 0.2 kcal per minute (Guyton, 1966) or about 0.46 liter per minute (for a latent heat of vaporization of 0.58 kcal per gram of water, and 0.75 gram of water vapor per liter). The maximum rate corresponds to sweat production at a rate of 25 grams per minute (Guyton, 1966) or about 33 liters of water vapor per minute. The above latent heat rate of 1.0 kcal per minute corresponds to 2.3 liters per minute.

The minimum volume of space per person V can then be derived as follows, assuming that the initial temperature is 20°C (68°F) and the relative humidity is 50 percent (8.7 mm Hg partial pressure):

- *Oxygen limitation.*

  $0.208V - 0.21t = 0.18V$, or V/t = 7.5 liters/minute-person

- *Carbon dioxide limitation.*

  $0.0004V + 0.21t = 0.03V$, or V/t = 7.1 liters/minute-person

- *Heat limitation.*

  $0.2 = 1.2V \times 0.24 \times (33\text{-}20)/t$, or V/t = 53.4 liters/minute-person

- *Humidity limitation.*

  $0.011V + 2.3t = 0.049V$, or V/t = 60.8 liters/minute-person

Thus, temperature and humidity considerations are more important than the oxygen or carbon dioxide limitations (Duff and Husband, 1974).

The largest of the above minima is 60.8 liters per minute per person, or 2.1 cubic feet per minute per person. Then

$$V_{total} = (2.1t + 2.65) N$$

where N is the number of people which can safely remain in a haven of volume V_{total} (cubic feet) for a duration of t (minutes).

Thus, the following table can be developed:

Duration of stay in haven (minutes)	Volume per person (cubic feet)	Area per person[a] (square feet)	Number of people in 100 square feet
5	13	1.6	60
10	24	3.0	33
15	34	4.3	24
30	66	8.2	12
60	1281	6.1	6

[a] Typical room height of 8 feet assumed.

For reference, 3 square feet per person is considered a "jam point" where all movement by the occupants is effectively stopped (NFPA, 1985, pp. 250, 305) and 7 square feet per person prevents normal walking, creating a "shuffle." A normal occupant load for an assembly area of concentrated use without fixed seats is 7 square feet per person, and the minimum normal occupant load is 5 square feet per person (NFPA, 1985, pp. 247, 303).

Two exits should be provided for havens, preferably at opposite ends of the room (NFPA, 1985, p. 253).

Although a haven should have tight-fitting doors and windows to prevent infiltration of outside air, the effect of persons inside a haven will be to provide a small degree of internal pressure (because of temperature increase and generation of water vapor) and net outflow of air from the haven.

REFERENCES

ACGIH, 1986: *Threshold Limit Values and Biological Exposure Indices*, American Conference of Governmental Industrial Hygienists, Cincinnati, OH, 7, 12.

Baumeister, T., et al., 1987: *Marks' Standard Handbook for Mechanical Engineers*, 8th Ed., McGraw-Hill, New York, 1296.

Chemical Industries Association, 1979: *An Approach to the Categorization of Process Plant Hazard and Control Building Design*, Chemical Industries Association, London, 25.

Duff, G. M. S., and P. Husband, 1978: Emergency planning. *Int. Loss Prev. Symp. 1st (The Hague/Delft)*, 121.

Guyton, A. C., 1966: *Textbook of Medical Physiology*, 3rd Ed., W. B. Saunders Company, Philadelphia, 554, 566, 979, 987, 989.

NFPA, 1985: Life safety code handbook. *National Fire Code*, NFC 101, National Fire Protection Association, Boston.

APPENDIX F

SOURCES TO VAPOR-MITIGATION EQUIPMENT VENDORS

	Types of equipment (page numbers indicated)				
Reference	Detectors	Flares	Scrubbers	Stacks	Spill alert
Thomas Register of Products and Services (1986) Thomas Publ. Co., New York	3-5056	5-7017	10-16705	10-15701	--
Hydrocarbon Processing Catalog and Directory, (1987) Gulf Publ. Co., Houston	35, 47	52	76	41	--
Chemical Engineering Catalog (1984) Reinhold, New York	10-1, 10-19	12-8	3-26, 12-11	12-13	--
Chemical Engineering Equipment Buyer's Guide (1987) McGraw-Hill, New York	425, 434, 444-447	463	515	463	--
Plant Engineering Directory (1987) Cahners Publ. Co., Des Plaines, Iowa	20, 74	--	18	43	--

Reference	Types of equipment (page numbers indicated)				
	Detectors	*Flares*	*Scrubbers*	*Stacks*	*Spill alert*
Industrial Hygiene News Catalog and Buyer's Guide (1984) Rimbach Publ. Co., Pittsburgh	35, 52, 61	--	--	--	101
Best's Safety Directory (1987) A. M. Best Co., Oldwick, N.J.	690-746	760	623	--	696
Occupational Safety & Health Sourcebook (1986) Medical Publication Inc., Waco, Texas	34, 44, 52	--	--	--	67

INDEX

Absorbers in emergency release, 46
Absorption/scattering sensors, 79
Accidents
    estimating consequences of, 120-121
    estimating probability of, 122
Administrative controls in flammable
        materials processing plant, 93
Adsorption of vented vapors, 46-47
Air curtains, 91
Alarms
    and air contaminant or leak
        detectors, 84-85
    in emergency flare systems, 42
    emergency scrubbers and, 35
    in flame arrester, versus "flame-
        holding" at arrester, 45
    vapor or "fume", 102
Alerting local authorities and public,
        113-116
    information to be communicated, 116
    roles and lines of communication, 115
    systems for, 114-115
Ammonia, anhydrous, handling, 19, 20
Ammonia vapor cloud control by water
        spray, 89
Anticipated instantaneous load, 31
Antifreeze protection, catchtank, 44
Audits and inspections, external, 67-68

"Base case" drawing of unloading
        process, 1-2

Blast-resistant buildings, explosion
        hazards and, 21
Breakaway couplings, 51

Carbon dioxide limitation in havens,
        139
Catalytic sensors, 78
Catchtank(s)
    with demister pads, 42
    design of, 133-135
    disposal of liquid from, 44-45
    operation and safety considerations
        in, 44
    for vapor-liquid separation, 43-45
Cause-consequence analysis, 122
Chemical burns due to vapor clouds, 4
Chemical reaction sensors, 78-79
Chemical substitution for inherent
        safety, 17
Chlorine Institute kits, 74
Chlorine Manual, The, 83
Combustible materials, flares to burn,
        40-43
Combustion sensors, 78
Communication
    to off-site populations
        alerting roles and lines of, 115
        information for, 116
    between process operators and
        supervisors, 64
    of warnings to personnel. 102-103

Community Awareness and Emergency
Response program (CAER), 115
Computers, use of
in automation of emergency abort
systems, 49
in emergency communications, 102
with process safety systems, 29
to streamline emergency response, 85
Containment
double, 25, 57
of spill, 57-60
Containment failure causes, 123-127
system failure events, 7
Corrosion testing, 25-26
Costs of major releases of fluids, 3
Countermeasures, mitigation through,
87-97
avoidance of factors that aggravate
vaporization, 96
liquid release, 87, 93-97
vapor release, 87, 88-93
Covers for spills, 94-96, 97
Curbs for spills, 59

Depressuring system, emergency, 55
Detection, early, 77-85
alarm systems and, 84-85
detectors and sensors and, 77-81
by personnel, 81-84
Dikes for spills, 58-60
Dilution
of liquid spills, 94
of released vapor, 19-20
Discharge of liquid downhill, 32-33
Discharge piping to catchtank, 44
Discharge systems, 31
Dispersion, gas, 38, 39
Double containment, 57
Double-walled piping, 25
Dump system, 48

Electrical sensors, 78
Emergency depressuring system, 55
Emergency plans and procedures
abnormal operation or, 64-67
abort systems, 47-49
features of (EPA and National
Response Team), 109-110

isolation of leak/break, 49-53
on-site response, 101-110
training and drills, 108-110
shutdown equipment and procedures,
103
transfer of fluids, 53-56
to reduce driving pressure, 54-55
to reduce inventory, 56
Emergency relief treatment systems, 33-
47
absorbers, 46
active scrubbers, 34-35
adsorbers, 46-47
catchtanks for vapor-liquid
separation, 43-45
condensers, 47
flares, 40-43
incinerators, 45-46
passive scrubbers, 36-37
stacks, 37-40
Emergency response, on-site, *see* On-
site emergency response
Enclosures and walls for containment,
57-58
Engineering design approaches to
mitigation, 23-60
plant physical integrity, 23, 24-26
process design features for
emergency control, 23-24, 33-
56
process integrity, 23, 27-33
spill containment, 57-60
Equipment, underpressuring of, 126
Equipment testing, 68-70
Escape from vapor cloud, 106
Event-free analysis, 122
Explosion
hazards of, and blast resistant
buildings, 21
internal, failure due to, 125-126
vapor cloud, 9

Facility, hazards of change to, 71-73
"Facility change review", 72-73
Fault-free analysis, 122
Fire(s)
due to vapor clouds, 8-9
flash, 9
vessel weakening due to, 55

Fire-alarm systems as vapor or "fume" alarms, 102
Fire extinguishers, 74-75
Fire protection foam makers, 95
Fire-retardant clothing, 107
Fireball, 9
"Flame-holding" at arrester versus alarm in arrester, 45
Flammable materials, emergency plans for, 109
Flammable-toxic vapor clouds, definition, 4
Flammable vapor clouds, definition, 3-4
Flares to burn combustible materials, 40-43
  safety considerations in, 42-43
Flashing, adiabatic, 5, 6
Flashing liquid
  and nonflashing, definition, 5
  special considerations for releases of, 5-6
Flow rate(s)
  of flammable gas in stack, 41-42
  mass, 66
  range of, and flares, 41
Flow restriction, 6
Foam covers for spills, 95
Fog correlations between vapor concentration, relative humidity, and heat of solution, 131-132
Freeze protection, 32
Freezing to stop leak, 6, 74-75
Freezing point data, 75

"Guard tank" ("scrub tank"), 36-37

Havens, 104-106, 137-140
  being infiltrated with contaminant from outside, 106
  capacity of, 137-140
  duration of stay in, 140
  entry into, 106
Hazard identification methods, 120
Hazardous materials, properties of, 129-130
Heat exchangers, shell-and-tube, 47
Heat in havens
  generation of, 138

  limitation on, 139
Highways, drifting of vapor clouds across, 4
"Holding" flame, by arrester, 42
Humidity limitation in havens, 139

Ignition of hazardous material, deliberate, 91-92
Ignition source control, 92-93
"Immediately Dangerous to Life & Health' (IDLH) concentrations established by National Institute for Occupational Safety and Health (NIOSH), 16
Imperfections in equipment, causes of, 68-69
Impoundments for spills, 59
Incinerators, emergency relief and, 45-46
Inherently safer plants, 11, 15-21
  chemical substitution and, 17
  dilution and, 19-20
  inventory reduction and, 15-16
  process modification and, 11, 17-20
  refrigerated storage and, 18-19
  siting considerations and, 20-21
Insulation material, choosing, 26
Inventory reduction for inherent safety, 15-16
Isolation of chemical process
  devices for
    breakaway couplings, 51
    inspection and testing of, 52-53
    pumps for, 51
    remote, 52
    valves for, 50-51
  by distance, 20-21

Jet entrainment
  of air, 38
  of flashing liquid, 6

Knock-out drum, *see* Catchtank(s)

"Leak before break", ensuring, 26
Liquid covers for spills, 94-95

Liquid release countermeasures, 93-97
   covers, 94-96, 97
   dilution, 94
   foams, 95
   neutralization, 94
   solids, 96
"Load-leveling" mechanism, volume
     system as, 34
Loss-of-containment causes, 123-127
   system failure events, 7
Low-temperature embrittlement, 6

Maintenance programs for process
     equipment, 70-71
Materials of plant construction, 25-26
Mechanical reaction forces in piping
     downstream, 33
Medical treatment for exposed persons,
   108
Mist eliminators, 43
Mist formation, 5
"Mitigation", defined, 2
Mitigation measures, selection of, 117-
   122
   methods for estimating consequences
     of accidents, 120-121
   methods for estimating probability of
     accidents, 122
   methods for hazard identification,
     120
   risk analysis, 117-118, 119-120
   sequence, 118
Moisture in havens, generation of, 139

Neutralization of spilled liquid, 94
Nitroglycerine production, 18

Odor warning properties, 82-83
On-site emergency response, 101-110
   communications in, 102-103
   escape from vapor cloud in, 106
   havens in, 104-105
   medical treatment in, 108
   personal protective equipment in,
     106-108
   plans, procedures, training, and drills
     in, 108-110

   shutdown equipment and procedures
     in, 103
   site evacuation in, 104
Operating Policies and Procedures
     program, 63-66

Patching, 73-74
Personal protective equipment, 106-108
Personnel
   communicating warnings to, 102-103
   release detection by, 81-84
     color or fog, 83-84
     odor warning properties, 82-83
     olfactory fatigue, 83
Plant conditions, deviations in, beyond
     design limits, 124-127
Plant physical integrity, 23, 24-26
   design practices and, 24-25
   inspection and testing during
     construction and, 26
   materials of construction and, 25-26
Plants, inherently safer, *see* Inherently
     safer plants
Plugging materials, segregating, 32
Plume rise
   flashing liquid, 6
   gas, stacks and, 38
Pool boiling, 5
Pressure relief systems, 30-33
Process change, hazards of, 71-73
Process control systems, 29
Process design features for emergency
     control, *see* Emergency plans
     and procedures; Emergency
     relief treatment systems
Process integrity, engineering design
     and, 23, 27-33
   headers, 31-33
   identification of reactants and
     solvents, 27-28
   limits on operating conditions, 28-29
   pressure relief systems, 30-33
Process modification for inherently
     safer plants, 11, 17-20
Process safety management approaches
     to mitigation, 63-76
   audits and inspections, 67-68
   equipment testing, 68-70

Process safety management approaches
to mitigation (Continued)
and hazards of modifications and
changes, in process or facility,
71-73
maintenance programs, 70-71
methods for stopping leak, 73-75
freezing, 74-75
patching, 73-74
operating policies and procedures,
63-66
security, 75-76
training for vapor release prevention
and control, 66-67
Property losses, vapors causing, 4
Pumps used for isolation, 51
Pump rate, stack, 41

Quench fluid quantity necessary for
catchtank, 133-134
Quench systems, 48-49

Reactants and solvents, identification
of, 27-28
Refrigerated storage and inherently
safer plants, 18-19
Relief headers, 31-33
Relief valves, 30-33
pilot-operated, 32
Remote ignition devices, 92
Remote isolation, 52
Respiration, havens and, 138
Respirators, 106, 107
Risk analysis, 117-118, 119-120
sequence, 118
Rupture disk, 30-31
or safety valve, discharge piping and,
44

Safe work practice, 71
Safety, inherent, *see* Inherently safer
plants
Safety review, pre-start-up, 73
Scrubbers
active, 34-35
passive, 36-37
Scrubbing liquid, 34, 37

Security, industrial, 75-76
Seismic effects, 25
Self-contained breathing apparatus
(SCBA), 107-108
Sensors, 77-81
positioning of, 81
response time of, 79-81
types of, 78-79
Shutdown devices, 28, 29
Shutdown equipment and procedures,
emergency, 103
Shutdown systems, leak-sensing, 49
Site evacuation, 104
Siting considerations and inherently
safer plants, 20-21
Solids as covers for spills, 96
Solvents and reactants, identification of,
27-28
Spill containment, 57-60
Stack flame arrester, 42
Stacks, 37-40
advantages of, over low-elevation
releases, 28-29
flare, 41
operation and safety considerations
of, 39
Steam curtains, 90-91
Subsidence, 25

Telephone dialing, automatic, for
communicating warnings, 102
Temperature, failure due to
metal, 126-127
process, 125
Toxic effects of vapor releases, 8
Toxic gases and vapors, emergency
plans and, 109
Toxic spill alerting systems, 114
Toxic vapor clouds, definition, 4
Transfer of fluids
to reduce driving pressure, 54-55
to reduce inventory available for
release, 56
Trenches for spills, 59
Turbulence and vapor cloud explosion, 9
Two-phase (vapor/liquid) flow, 5, 43-45

Unloading process, "base case" drawing
of, 1-2

Valves
  check, 51
  to drain or vent down equipment, 56
  excess-flow, 51
  for isolation, automatic and manual,
    50
Vapor cloud(s)
  escape from, 106
  explosion of, 9
  types of, 3-4
Vapor cloud hazards, emergency plans
    for mitigation of, 107
Vapor concentration and relative
    humidity and heat of solution,
    fog correlation and, 131-132
Vapor/cover gas, transfer of, to reduce
    driving pressure, 54-55
Vapor-liquid (two-phase) flow, 5
  catchtanks for separation of, 43-45
Vapor-mitigation equipment vendors,
    141-142
Vapor pressure of liquid at elevated
    temperature, 54-55
Vapor releases
  actions taken in design stage and, 11
  active and passive approaches to, 10-
    11
  analysis of need for mitigation in, 10
  causes of, 7
  countermeasures to, 87, 88-93
    air curtains, 91
    deliberate ignition, 91-92
    ignition source control, 92-93
    steam curtains, 90-91

    water curtains, 89-90
    water sprays, 88-89
  emergency-response actions and, 11,
    12
  engineering design of plant and
    process and, 4-6
  forms of, 4-6
  hazard of accidental, 3
  inherently safer plants and, 11
  possible consequences of, 8-9
    explosions, 9
    fires, 8-9
    toxic effects, 8
  plant and process integrity and, 10
  training for prevention and control
    of, 66-67
Vaporization, avoidance of factors that
    aggravate, 96
Vapors
  condensing and noncondensing, 4-5
  dense, 6
Vessel weakening due to fire, 55
Visual sensors, 79
Vortex separator design, 133, 135

Wake effects, 39
Warning, *see* Alarms
Water curtains, 89-90
Water seals, flares and 42, 43
Water sprays for vapors, 88-89
Wind loading, 25
Wind speed, upper-air, 38-39